**Globally
Optimal Design**

GLOBALLY
OPTIMAL DESIGN

DOUGLASS J. WILDE
Design Division, Mechanical Engineering
Stanford University

A WILEY-INTERSCIENCE PUBLICATION

JOHN WILEY & SONS, New York ● Chichester ● Brisbane ● Toronto

Library of Congress Cataloging in Publication Data

Wilde, Douglass J.
  Globally optimal design.

  "A Wiley-Interscience publication."
  Includes bibliographies and index.
  1. Engineering design.   2. Mathematical
optimization.   I. Title.

TA174.W46   620′.004′2   78-2933
ISBN 0-471-03898-9

Printed in the United States of America

10 9 8 7 6 5 4 3 2 1

TO MY DISTINGUISHED RESEARCH STUDENTS

Vargaftig, Thomas, Avriel, Passy, Beamer, Blau, Reklaitis, Williams, Sanchez, Rijckaert, Barthés, Atherton, Van Remortel, Shapiro, Mancini, Lidor, McNeill, Ait-Ali, Papalambros, Human

Homines, dum docent, discunt.
While teaching, one learns.

Seneca (c. 63 A.D.)

Simab said:

> *I shall sell the Book of Wisdom for a hundred pieces of gold,
> and some will say that it is cheap.*

Yunus Marmar replied:

> *And I shall give away the key to understanding it,
> and none will take it, even free of charge.*

Idries Shah, *Thinkers of the East*

Preface

This book applies a new optimization theory to engineering design in a way that is also new. Past optimization theory, developed for operations research, has had little to offer the engineering designer, whose problems rarely satisfy the restrictions necessary for making nonlinear programming algorithms work. Until now, designers brave enough to use optimization at all have been forced to use numerical techniques with no guarantee of finding the right answer, even after long and expensive computer runs.

How is it that an experienced designer can often, after only a few simple calculations, produce a better design than can a computer-supported optimization code? The answer is that the key structure in design problems is different from that on which conventional optimization theory is based. Yet this structure is so simple that experienced designers sense it and use it intuitively. By exposing this structure and putting it on a rigorous but clear mathematical basis, this book points the way to design procedures, simple enough for a handheld calculator, that guarantee a design numerically indistinguishable from the global optimum.

Global is the key word. This book shows why many common design problems, from reservoirs to refrigerators, have multiple local optima, as well as false optima, that make conventional optimization schemes risky. Consequently, only procedures guaranteeing global optimality are presented. Since many of these are novel, the book will also interest operations analysts and applied mathematicians wishing to push back the limitations of optimization theory.

This is a text for an elective graduate course for engineers and operations analysts developed in Stanford University's Design Division in 1973. To be admitted, students must have completed an undergraduate engineering design course, and all at Stanford are required to apply the material to an individual design project of their own choosing. One purpose of this book is to strengthen the design profession by making possible the introduction of similar courses in other universities. Being applicable to all engineering disciplines, such a course could originate anywhere in the college of engineering where there are faculty willing and able to teach it. In many schools this will be the operations research department, provided the instructor has enough engineering background to cope with the unavoidable jargon and empiricism of the design specialists.

Three lecture hours a week for a semester should be enough to cover the material. If only a quarter is available, Chapters 7 and 8 may have to be included in another course or omitted. Chapter 1 defines optimal design, giving historical background on both optimization and design. Specifications and the powerful simplifying concept of monotonicity are discussed in Chapter 2. Chapter 3 shows how constructing estimates and bounds eliminates unnecessary computation. This idea is extended in Chapter 4 to constrained problems, determining which inequalities must be satisfied as equalities at the optimum. Chapter 5 copes with profit, whose negative terms confound the geometric programming based ideas of the preceding two chapters. Chapter 6 demonstrates how negative signs in constraints, especially in thermodesign, generate multiple local optima; it then develops globally optimal solution procedures. The more exotic transcendental forms arising in chemical reaction and mass diffusion problems are dealt with in Chapter 7. How to handle standard sizes, indivisible variables, and logical problems expressed in natural language is discussed in Chapter 8; a computation scheme called "lexical arithmetic" is also introduced. The entire work is located in Chapter 9 within the framework of engineering design.

Perhaps a decade of research, carried out by the chemical engineering, operations research, and mechanical engineering students named in the dedication and supported mainly by the U.S. National Science Foundation, preceded the writing of this book. The book, which is a sabbatical-leave project underwritten principally by Stanford University, owes its existence also to the hospitality of the Pontifícia Universidade Católica do Rio de Janeiro in Brazil and to the United Nations Development Program in Rangoon, Burma. There, and in hotels, huts, and houseboats from Kathmandu to Cuzco, this book was brought into being.

DOUGLASS J. WILDE

Stanford, California
April 1978

Contents

Globally
Optimal Design

CHAPTER 1

Optimization and Design

*The value of a principle is the number of things it
will explain.*

Ralph Waldo Emerson (1867)

Design engineers, after conceiving a device, must choose parts both economical and safe. From experience they learn to recognize good proportions and critical restrictions, so that their preliminary designs, if not already the best attainable, need little improvement. This ability, appearing intuitive or even magical to the uninitiated, is the subject of this book.

The vehicle for explaining and understanding this talent is a young branch of applied mathematics known as "Optimization Theory," a science that studies the best. But this is not a description of optimization theory so much as it is an account of those principles of optimal design presently capable of precise description. Many of these concepts, being based on previously unrecognized mathematical properties of design problems, are in fact novel contributions to optimization theory as well as to engineering design.

The style is intended to please engineers without offending the mathematically inclined, who wish to apply their disciplines to design. As in life, where instinct precedes wisdom, a realistic problem introduces each topic, and like the experienced designer, the reader must understand the technology before confronting any mathematics. Each problem is modeled as optimizing an objective function subject to constraints. But instead of trying to solve the problem numerically with general purpose optimum seeking codes, the method is to deduce abstract principles for finding the optimum design with as little computation as possible. The goal is not only a simple design procedure for the specific problem, but also the discovery of rigorous ways to recognize an optimal design.

Although laymen may view engineers as paragons of precision, mathematicians more often see even good designers as exercising excessively Ambrose Bierce's perogatives of genius (1911): "To know without having learned; to draw just conclusions from unknown premises; to discern the soul of things." Thus many designers may recognize, in the principles derived here, rules they use frequently, if unconsciously, in their own practice. Mathematical rigor is employed sparingly, no more than needed, to prove these design rules and circumscribe their applicability. Although codification may diminish design's romance, it will strengthen the profession and increase its accessibility. Samuel Johnson said (1784), "Genius is nothing more than the use of tools, but there must be tools for it to use."

The way a design problem is formulated strongly affects its ease of solution. This may explain why elaborate computer studies all too often yield little or no improvement over a design made by a mathematically naive engineer familiar with the technology. Hence previously overlooked mathematical properties of design problems should interest not only engineers, but also all optimizers, be they operations analysts, computer scientists, mathematicians, economists, or businessmen.

Operating problems have generated an optimization theory of only limited worth to designers, who without knowing it face situations that are nonlinear, nonconvex, nonunimodal, noncontinuous, and consequently, at least to many an operations analyst, nonsolvable. This book copes with such problems which, although widespread in design, are not confined to it: specifically, engineering systems with objective and constraints, equality as well as inequality, composed of both algebraic and transcendental nonlinearities depending in part on discrete variables are optimized globally, not just locally. Since the properties making this possible may also occur in operating problems, the attention is justified of the entire optimization fraternity, even members who never encounter engineering design.

1.1. THE QUICK, THE CLEAN, AND THE SATISFYING

Before the large electronic computer and the attendant development of optimization theory, many design methods were characterized picturesquely as "quick and dirty." Quick meant easy to calculate; dirty, approximate. These procedures gave designs conservative in the sense of being workable (an optimizer would say "feasible") but not necessarily most economical. More recent formulation of design as an optimization problem lead to computer methods that, while no longer dirty, neither were any longer quick, requiring elaborate modeling and extensive, usually iterative, computation. Designers would, of course, prefer procedures both quick and clean.

A major goal of this work then is to expand the concepts, both of good design procedure and of good optimization technique, to include the quick and clean. Presently, the ideal quick design procedure uses one or more formulas, whereas a good clean optimization technique is often an iterative algorithm known to converge, albeit under special circumstances not always present in a given problem. For the design examples solved here, however, formulas would require oversimplification, whereas elaborate algorithms are not really needed. What results instead is a system of what might be called *conditional* formulas. Design then involves evaluating a formula, the results leading to one of several possibilities. After checking a small number of such cases, the designer finds one guaranteed to be globally optimum. Procedures free of loops and iterations are preferred, but where they cannot be avoided, bounds are constructed to tell how much improvement to expect if computations continue. This prevents unwarranted iteration. Thus the individual steps of a conditional procedure resemble design formulas, but the overall logic reduces unproductive computation and unnecessary modeling; information is requested only when needed. The examples show that, armed with a good conditional design procedure and a pocket calculator, the engineer can solve significant problems that otherwise would need an expensive computer that might not even give the right answer. Conditional methods also increase the size of problems solvable on large computers. As with the mathematical properties of design problems, the idea of conditional procedure may ultimately prove as useful to optimization theory as to design.

An attitude pervades these pages that, although novel neither to engineering nor to optimization, deserves emphasis. Euripides (410 B.C.) expressed it as: "To the wise, enough is abundance." In this vein, most engineers accept any design suitably close to optimal, preferring not to waste time pursuing absolute, even unattainable, perfection. Given the uncertainty of physical and economic data, this position is certainly reasonable; the difficulty is knowing when to stop looking. Consequently, the book shows how to construct bounds that tell when diminishing returns on computation and analysis are reached. This greatly simplifies things when quick rules generate nearly optimal designs, or when many designs give nearly the same performance. If we ignore this consideration, algorithms that generate needless iterations bring to mind an aphorism of Epicurus (c. 300 B.C.): "Nothing is enough for whom enough is too little."

1.2. PREVIEW

The first chapter, a brief history both of engineering design and of optimization theory, sets the stage. The last recapitulates the information presented and points to the future. The middle seven are organized by those characteristics of design problems lending themselves to optimization.

In this vein Chapter 2 deals with specifications; sometimes the vaguest fragment of information about design objectives can determine which specifications are critical at the optimum. This alone may set the values of most design variables, greatly simplifying the problem. Two techniques for identifying binding constraints, called partial optimization and monotonicity analysis, are applied to designing pressure vessels.

The hallowed engineering practice of estimation is discussed in Chapter 3. Here it is shown how cost bounds can be constructed for the power function nonlinearities peculiar to design. Upon these is based the theory of geometric programming that has greatly advanced the science of design in recent years. By constructing bounds the designer may find, as in the cofferdam example, when a quick estimate may be suitably close to optimal, making further computation unnecessary.

In Chapter 4 these bounding procedures are extended to problems with constraints that, like the objective of Chapter 3, are sums of power functions. A more advanced version of the monotonicity analysis of Chapter 2 is developed and used to determine which inequalities will be satisfied as strict equalities at the optimum. This leads to "conditional" design procedures that give the global optimum after very simple computations, as illustrated by a hydraulic cylinder design. A merchant fleet design problem shows how to apply these ideas to a moderately large problem with nine variables and eight inequality constraints, previously solved only by a computer code.

Profits, which introduce negative terms confounded by the earlier bounding procedures, are studied in Chapter 5. Since geometric programming can no longer be applied, other methods previously of secondary interest now become central, being the only available. Construction of bounds and finding the global optimum is, however, still possible.

Negative signs in the constraints are shown in Chapter 6 to generate multiple local optima and stationary points that mislead conventional methods not constructing bounds. An irrigation pipeline and reservoir system, a chemical reactor and cooler, and a refrigerated storage tank are designed by these methods. New ways of constructing bounds are demonstrated.

Chapter 7 treats exceptional systems with transcendental variables not fitting the power function form needed by previous methods, whose variables are now called "algebraic." Methods of the preceding chapters still apply to the algebraic variables, leaving a residue of transcendental variables that do not cancel out. Under certain easily verified circumstances, bounds can be constructed even in this case. A gaseous diffusion plant for separating isotopes is designed by these methods.

By making the design variables discrete rather than continuous, standard sizes might seem to invalidate many of these methods. On the contrary, Chapter 8 shows that many concepts can be extended to the discrete case. In fact, standard sizes simplify the pressure vessel, pipeline, and cofferdam examples. A harder problem of picking the best among over a thousand tolerance assignments is then solved handily with some accessible ideas from combinatorial optimization theory. This leads to a tabular branch and bound procedure, which is easily extended to cases where specifications are verbal logical statements instead of algebraic equations. A problem with safety rules, environmental laws, and marketing considerations joining economy as objectives and specifications is solved as an example. The chapter closes with a Holmesian murder mystery demonstrating the reduction of verbal statements to precise logical ones amenable to optimization.

The final chapter summarizes all these developments, putting them in perspective from the designers' as well as the optimizers' point of view. Possibilities for the future are also discussed.

The aims of this book are to make design more rational and optimization more practical. Is the designer any less an artist who becomes more a scientist?

1.3. FUNDAMENTAL DEFINITIONS

This book intends to help engineers find, from among all workable designs, that which is best. In ordinary English, the words "design" and "best" each have several meanings. To reduce confusion, these meanings will be discussed now in detail so that a narrower, precise construction can be placed on each word.

Human artifacts, such as arrowheads, clothing, buildings, or machines, reflect the thoughts and imagination of the *designers* who plan them. The *design* of such an object can be the detailed shapes, colors, materials, dimensions, and arrangement of its parts. Alternately, a design may involve the performance expected from certain components: speed, life, efficiency, or cost. Roughly speaking, a design represents the thinking that goes into an artifact.

There is a spectrum of design: professions range from fine artist, motivated by pure aesthetics, to engineer, for whom appearance is overshadowed by utility, with fashion designers and interior decorators somewhere between. Although aesthetics should rarely be ignored entirely, even by the engineer, this book concentrates on the utilitarian aspects of design. It speaks to the engineer rather than the artist, and the design principles developed here are scientific rather than aesthetic.

As intended here, scientific or rational design excludes not only aesthetic considerations, but also visual ones not lending themselves to mathematical expression. This may seem to ignore those important aspects of design expressed in the drawings, sketches, plans, and patterns communicating a design to the builders. But because most of the information in an engineering drawing is geometric and numerical, it can in principle be expressed quantitatively as a set of equations, verbal instructions, or computer codes for purposes of analysis. Thus the mathematical formulations used here are to engineering drawings as algebra is to geometry—an equivalent language for representing the same physical object. Ideally, designers should be skilled in mathematical as well as visual thinking, using whichever is more convenient in a given situation. While recognizing the importance of the visual side of design, we concentrate in this book on the mathematics.

To design something intelligently, a designer must know what its purpose is in the face of any restrictions placed on it. These goals and limitations are expressed rationally as specifications or constraints; for example: "The vessel shall withstand 10 atmospheres internal pressure and not exceed 150 centimeters in outside diameter." A design satisfying all constraints is said to be "workable" or "feasible," and the main task of the designer is to produce at least one feasible design.

When constraints are relatively unrestrictive, the designer may be able to choose between many different designs, all feasible. If the feasible designs can be evaluated in some quantitative manner, say by estimating their costs, they can be compared. It then becomes meaningful to speak of one design's being better or worse than another. This book shows how to find a rational design that is either best of all or very close to it.

For brevity as well as precision, the expression "best feasible rational design" will be replaced by *optimal* design or simply optimum. Although strictly speaking they are synonyms for best, the adjective optimal and the noun optimum are technical terms connoting a mathematical, quantitative context. Optimum was first used in this sense early in the 18th century by the philosopher-mathematician G. W. Leibniz, whose invention of differential calculus for finding maxima and minima of mathematical functions carried into his theological speculations in a manner relevant even to a tough-minded designer of today. He expounded the theory that this world is optimum, which has been translated into English as "best of all possible." Thus feasibility is an essential overtone of the word "optimum," for what is best may not be possible.

The relation of feasibility to optimality was widely overlooked in the 18th century, leading to misunderstandings worthy of discussion in the 20th, where similar confusion is emerging. Leibniz did not think this optimum world very good; with his theory he in fact tried to reconcile the world's obvious defects with his Christian notion of a wise, powerful, and benevolent creator (or, in this context, designer). He died too soon after publishing his *Theodacy: Essays on the Goodness of God, the Freedom of Man, and the Origin of Evil* to keep his followers from overlooking feasible to concentrate on best, distorting his ideas into what came to be called "Philosophical Optimism"—the belief that everything that happens, good or bad, is for the best and must be endured stoically rather than confronted and changed. Originally a technical term for adherents to philosophical optimism, the word "optimist" came to mean one who sees only the bright side of things by unrealistically ignoring any unpleasantness. Even philosophical is often taken popularly to mean passive acceptance of trying circumstances. The 20th century equivalent of the philosophical optimist is the designer who accepts a bad design just because it is optimal, that is, the best available, without seeking to push back the constraints.

Optimum also connotes a *quantitative* measure of desirability. Leibniz derived his word optimum from the Latin *optimus*, meaning best. An older meaning of *optimus*, however, was most, the word coming from *Ops*, the name of the Sabine goddess of fertility and agricultural abundance. The ancient root of optimum, in carrying the notion of quality measured quantitatively, therefore harmonizes with the contemporary technical meaning.

The *optimal design* of a device is the feasible plan that makes it as good as possible according to some quantitative measure of effectiveness. With capital letters, *Optimal Design* is defined as the applied science of finding optimal designs.

Ops, in Roman mythology the wife of the god of time Saturn and mother of the god of power Jupiter, would be a fitting patroness for the fledgling science of Optimal Design. With the cornucopia, the fabled horn of plenty under one arm, Ops holds in her other hand scales symbolizing measurement and decision. Her name appears not only in cornucopia itself, but also in modern words for the fruits of labor—opus and opera—as well as in operation, the process of producing them. Operations research, a close relative of optimal design soon to be described, also contains the name of the goddess of abundance. During the festival of Ops in December, the Romans opened the graneries, a custom anticipating the

Figure 1-1   The Goddess Ops

exchange of presents at Christmas. May Ops in her generosity grant gifts of abundance, time, and power to all who practice optimal design.

Now that optimal design has been defined, it is time to write of its antecedents in engineering design and in optimization theory, both being on the periphery of better known disciplines. The next section treats engineering design, elements of which appear in all branches of engineering. Following this is a discussion of optimization theory, which may be regarded as pertinent to many branches of mathematics while belonging to none of them. The intersection of engineering design with optimization theory generates the discipline of optimal design.

1.4.   ENGINEERING DESIGN

Since optimal design may be viewed as optimization theory applied to engineering design, an understanding of engineering design, its history, tribulations, and future potential is in order. Engineering is organized academically and professionally by disciplines, the major ones corresponding roughly to the main scientific principles being applied: civil, chemical, electrical, materials, mechanical, and industrial. More specialized groups have formed using specific industries or devices as unifying themes: e.g., architectural, agricultural, aerospace, electronic, lubrication, control, and plastics engineering. Each discipline and specialty involves some design, along with other things engineers do: administration, manufacturing, troubleshooting, and research. For convenience, the aggregation of the design functions from all engineering disciplines is here called *engineering design*.

Although by this definition there are many engineering designers, there is no unified profession called engineering design. Engineering designers, also called design engineers, would if asked probably claim identity with their organized professions, speaking of themselves for example as machine designers or naval architects. Indeed, most of their expertise concerns the detailed technology of their specialty rather than how it is used in design. This diffuse distribution of the design function has impeded development and perfection of design techniques common to all specialties. Merely by viewing engineering design as a unity, this book hopes to encourage its refinement throughout the engineering profession.

The wide diffusion of engineering design has produced educational as well as professional difficulties. In its youth, engineering education had a large measure of technology complementing the scientific fundamentals taught to students. Aspiring engineers learned in detail how the devices and systems of their discipline functioned, as well as how to operate and design them for specific applications. Just before graduating, they took one or more appropriate design courses: machine design for the mechanical engineers, network synthesis for the electrical engineers, process design for the chemical engineers, structural design for the civil engineers. These design courses had little in common, especially as decades passed and technologies diverged. By the middle of the 20th century, the range of technology within each single discipline had grown so large that most students could expect to use in their careers very little of the knowledge acquired in design courses. These became repositories of inherently obsolescent information of but small relevance to a student's future. With irrelevance came dissatisfaction and inevitable reform.

By the 1960s the reaction against evanescent technological details had brought their wholesale purge from most engineering curricula. Although the increased role of scientific fundamentals for the most part strengthened engineering, it also caused an unfortunate deemphasis of design because of the latter's close association with technology. The baby, design, had been thrown out with the bath water, technology.

At this writing, counter reformation is in the wind, for it soon became apparent that the younger engineers, while more knowledgeable in basic science than their senior colleagues, were too often awkward with design problems. Recognizing this, the Engineers' Council for Professional Development (ECPD) placed a lower bound on the amount of design or synthesis that an American curriculum can have before losing its academic accreditation. Most colleges comply, under this pressure, by reintroducing the smallest portion of technological style design courses allowed by the accreditation team. A better way to teach engineering design might be to divorce it from technology and focus on what every engineer, regardless of specialty, needs to know about it.

Optimal design is one component of engineering design that, standing free of technological details, cuts across all engineering specialties. Others —graphic communication, visual thinking, engineering geometry, human factors, engineering case study, and creativity—deserve brief discussion here to place optimal design in perspective.

Graphic communication is a modern version of the old engineering drawing courses, deemphasizing tee-square rigor in favor of careful free-hand sketching and computer graphics. Visual thinking awakens the aesthetic, image-producing parts of the brain that have atrophied from the necessarily literal, rational, and mathematical education of the engineer. Engineering geometry is an advanced version of descriptive geometry dealing not only with detecting interference in three dimensions, but also with those portions of algebraic and differential geometry, notably curves, surfaces, and motions in hyperspace, of interest to designers. Human factors relates human biology and psychology to design. Engineering case study, using methods widely employed in medical, business, and law schools, prepares students for the imprecise and ambiguous world of engineering practice without burdening them with details of technology. Creativity stimulates the inventive imagination that generates novel concepts to satisfy new needs. All these subjects are often integrated in a long-term project course in which individuals or teams actually design, build, and test engineering devices.

Although such a curriculum may remain rare, practicing engineering designers, whether in or out of the university, benefit from mastering these subjects. For the best engineers, learning does not stop with graduation. This book, whether as an academic text or as a reference for the self educated, is intended to advance engineering design in all its specialized forms.

Optimal design fits in with these other elements of engineering design. Every device or system must have the sizes of its components specified exactly. "Sizing," as designers call procedures for doing this, is the intellectual ancestor of optimal design. The sizing methods that fill the engineering handbooks often are formulas proven, in simple situations, to optimize some reasonable quantity such as weight or cost. Learned by the designer along with the technology, these sizing techniques do not always guarantee optimized designs. For devices of even mild complexity, no sizing methods are available. Optimal design shows designers how to find either optimized sizes and characteristics in specific situations or else general formulas and procedures where none now exist. Optimized sizing procedures become part of the appropriate technology, while optimal design remains the intellectual framework for discovering and rigorously proving new sizing methods.

This completes the relating of optimal design to the rest of engineering design. It remains to discuss optimization theory, whence come many of the principles of optimal design.

1.5.   OPTIMIZATION THEORY

The process of making a desired quantity maximum, or an undesired one minimum, is known as *optimization*. That body of mathematics dealing with the properties of maxima and minima, as well as how to find them numerically, has come to be called optimization theory. Optimal design can be viewed as the optimization of those mathematical functions found in engineering design problems. Thus optimal design can immediately use any existing optimization theory dealing with the functions of engineering design. By the same token, many new developments in optimal design represent extensions of existing optimization theory.

Only recently has optimization theory accumulated enough content to qualify as even a minor branch of mathematics, although constructions involving optima go back at least to the ancient Greeks. Euclid's axiom that the shortest distance between two points is a straight line must certainly have been known to the Pharaoh's surveyors in the Nile delta. A more sophisticated principle of optimal design, discovered empirically and remaining unproven for millenia, was stated in a literary context in Virgil's legendary account, in his *Aeneid*, of the founding of Carthage. The gods had granted Queen Dido as much land as she could enclose with a bull's hide thong. She arranged the thong in a semicircle, its end against the Mediterranean Sea, thus taking the maximum area possible. This early urban-design principle, which gives the shortest wall perimeter for a given city area, can be seen today where the semicircular Ringstrasse boulevard, its ends on the Danube canal, stands on the foundations of the walls that guarded old Vienna. Here optimal design anticipated optimization theory, for it was not until the 18th century that mathematicians proved that a circle encloses the greatest area for a given perimeter in the plane.

The earliest example of optimization theory used to recognize and locate an optimum was in the 17th century, when Fermat observed from his numerical computations of curves that the rate of change of a function decreased to zero at a maximum or a minimum. Strengthened by the subsequent invention of differential calculus, this principle was used to solve such famous 18th century puzzles as the *isoperimetric problem* described in the preceding paragraph. Another was the *brachistochrone*

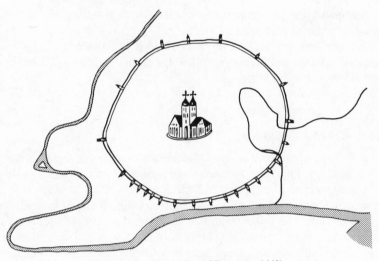

Figure 1-2   The walls of Vienna (c. 1440)

(Greek for "least time") problem, in which one seeks the vertical curve that a frictionless ball will traverse in minimum time. In solving this and similar optimization problems, Bernoulli, Euler, and Lagrange founded the calculus of variations, an early manifestation of optimization theory for finding functions minimizing an integral. Although the calculus of variations produced valuable contributions by Jacobi and Weierstrasse to 19th century mathematical analysis, not much optimization theory of interest to designers was available before invention of the high-speed electronic computer in the mid 20th century. An exception would be Courant's work on minimal surfaces, which has had some influence on contemporary architectural forms. But for the most part, engineering designers have used little optimization theory beyond the 18th century idea of setting first derivatives to zero in cases simple enough to yield closed-form solutions.

The lack of practical demand for optimization theory during that period naturally inhibited its development. Engineering concerned itself then with design rather than operation of existing equipment, and design problems of any complexity did not yield readily to optimization. But as industrial systems grew larger and more complex, and as more engineers became involved in manufacturing rather than design, a need emerged for better ways to manage operations. In response to this demand, the new profession of operations research, founded to apply science to military tactics during the Second World War, developed optimization techniques easily adapted to industrial operations. For, unlike most design problems, those involving operations are linear in that any inputs generate proportional outputs. This linearity, simplest of mathematical relations, makes possible precise analysis of operating systems too complicated for the unaided intelligence. Concurrent development of the high-speed electronic computer for the first time made practical the tedious but routine calculations needed to solve linear systems. Equally important, the simplex method for optimizing a linear function subject to linear inequality constraints—the linear programming problem—was published by G. Dantzig in 1947. The literally thousands of applications of linear programming, together with its extension to mildly nonlinear situations amenable to linear approximation, can be viewed as the start of modern optimization theory.

The success of linear programming moved operations researchers (also called "operations analysts") to study other situations lending themselves to optimization. Bellman expounded dynamic programming, a form of mathematical induction, as a way to optimize time-dependent, i.e., dynamic, systems of the kind arising in economic planning and automatic control. Pontryagin's topological updating of the calculus of variations, called the "maximum principle," removed an important theoretical barrier to the exploration of outer space. Many computer-oriented algorithms, or calculation schemes, were developed for iterative seeking of local optima, both with and without constraints. The esoteric mathematics of groups and combinatorics was brought to bear on problems called "integer programming" because of their indivisible variables. By 1970 all these accomplishments became an identifiable branch of applied mathematics, known as optimization theory to most engineers, and as mathematical programming to many operations analysts and computer scientists.

Although developed mainly for operating problems, these ingenious methods also found application to engineering design. Indeed, some optimization theory, such as the geometric programming of Zener and Duffin, was intended primarily for design engineers. This book shows how this and other special elements of optimization theory can advance engineering design.

On the other hand, some aspects of optimization theory are less likely to benefit design. Many design problems in the engineering literature are now, because of advances in optimization, formulated in terms of an objective function, say cost or weight, to be optimized subject to technological constraints by computer codes incorporating the latest algorithms. But even though methods devised for operations planning can in this way be applied to the very different problems of design, success or even efficiency is not guaranteed. Excessive computation, although perhaps justified in an isolated design study, is wasteful when the procedure must be used often. Worse, substituting calculation for analysis obscures any underlying design principles by which the engineer can improve future designs. Most wasteful of all is uncritical application of operations-oriented optimization methods, derived for circumstances absent in many design situations, which can generate designs that are not only not optimal, but may even be the worst possible! Extreme though that sounds, the engineering literature does contain depressing examples of optimization gone wrong.

This book focuses only on those parts of optimization theory pertinent to design engineers. Operations-oriented topics such as linear programming, and such control theory standbys as dynamic programming and the calculus of variations, are omitted. This leaves more space for geometric programming with its special power over the nonlinearities common to design, as well as for those "branch and bound" methods of integer programming especially suited for selecting standard sizes.

Much in this book is new to optimal design, and consequently to optimization theory. Special properties of design problems are identified so that appropriate optimization techniques can be devised to handle them. For example, the simple property of monotonicity, often yielding the optimal design after little or no calculation, is shown to be present in many design situations. Another novelty is the construction of simple functions bounding the original ones from below, permitting global, rather than merely local minimization. These lower bounding functions may also prove when a design, although not optimal, is good enough to accept without further computation. When the closeness of such a design to optimality can be established rigorously within acceptable limits, it is called a satisfactory design. A useful theory of *satisfactory design* is thus made available.

Fresh approaches to selecting parts available only in standard sizes are also presented. Much can be done by systematically manipulating constraint inequalities. For more complicated combinatorial design problems, a simple scorekeeping system called "lexical arithmetic" is developed that also solves interesting logical problems in design.

Further novelties concern the power functions of design, many of whose properties have been exploited by geometric programming. They are shown to possess other characteristics useful for numerical optimization. Other kinds of design nonlinearity, stemming from exponential functions or constraints based on physical conservation laws, are analyzed to give new optimization procedures. Generated by the needs of engineering design, all these developments, by extending the scope of optimization theory, will eventually contribute to operations analysis, management science, and mathematical economics.

What might be called the "designer's attitude," which permeates this book, may also widen the horizons of optimization theory. This point of view demands computation procedures powerful enough to be simple, which requires sensitivity to special mathematical properties overlooked by those not familiar with engineering, physics, or chemistry. Exposition of such properties should indicate new directions of research for optimization theory. The emphasis on simplicity, of not using a large computer where a manual calculator will do, encourages a taste for elegance rather than copious data processing.

1.6. POSITIVE PESSIMISM

This chapter has so far attempted to arouse interest in optimization by emphasizing its potential for finding better ways to design better things. Like most human activities, however, optimization can be misused, and it is only fair to conclude with words of caution.

There are at least four ways to misapply optimization. One is to use the word "optimal" as an adjective, not for the best possible, but merely for what the designer or vendor happens to prefer. Another occurs when a design is technically optimal, but the measure of effectiveness optimized, say cost, is very different from what the user wants, say performance. A more subtle abuse happens when the designer overlooks or ignores an important constraint, on service life or environmental impact for instance. Finally, perfectionistic obsession with the ideal optimum can be wasteful when several alternatives give results practically indistinguishable from the best.

Here are some remedies for these four abuses. To keep the word "optimal" from being degraded into advertising rhetoric, engineers should restrict the word to well-defined, quantitative situations where technical tests of optimality can be applied. As Seneca wrote almost 2000 years ago, "It is not goodness to be better than the worst," and it is misleading to speak, say, of an "optimal chair."

To prevent "optimal" from becoming a euphemism for "cheap," designers should make clear what objective is being optimized. Who wants a machine whose cost to the manufacturer has been minimized at the expense of reliability and operating economy? The Scots have a proverb appropriate for such conflicts of interest:

Twa gudes seldom meet.
What's gude for the plant is ill for the peat.

To keep "optimal" from meaning "shoddy," engineers, whether as designers, customers, or citizens, must be sure all constraints concerning safety, environmental impact, and service life have been included in any optimization study. Communities are beginning to crack down on "optimized" factories that pollute the biosphere or endanger the workers. Cervantes' question in *Don Quixote*, "Can one desire too much of a good thing?" has been posed to the designers of supertankers and supersonic transports.

To prevent "optimal" from connoting "impractical" or "academic," optimization theorists must avoid quibbling about easily computed "excellent" designs suitably close to optimal. Otherwise, "The best is the enemy of the good," in the words of Voltaire. Better to observe the Latin maxim *satis quod sufficit* (enough is as good as a feast).

The Welsh sum up these warnings with their poetic saying, "Talent without sense is a torch in folly's hand." Like most technical developments, optimal design can give good service if employed thoughtfully in appropriate situations. In the salty words of Benjamin Franklin,

He's the best physician that knows the worthlessness of the most medicines. *Poor Richard's Almanac* (1733)

1.7. STUDENT DESIGN PROJECTS

Graduate engineers of any specialty should have no difficulty with this book, which is based on a first year graduate course developed in the Design Division of Stanford's School of Engineering. The mathematics, a perhaps unconventional but transparent blend of differential calculus, vector and matrix notation, inequalities, and logic, is elementary enough to lure any engineering or science sophomore into studying the subject, but this has proven unwise. Lacking experience with design-oriented problem solving, such students mistake the mathematical shadow for the technological substance and are unable to apply what they learn. One had better be well grounded in engineering before tackling optimal design.

Indeed, the best way to learn optimal design is to design something with it. In the Stanford course, each student applies the theory to an individual design project, usually selected by the student from the engineering literature. This provides valuable experience in modeling, analyzing, simplifying, and computing (topics of value to any designer or optimizer) better learned by practice than discourse. The best of these student-generated problems are in the text either as examples or potential exercises. Encouraging students to find their own projects early in the course, before they have studied much of the theory, is strongly recommended. In this way students pose their questions before, not after, receiving the answers. Better still, they sometimes find answers where none existed before, thereby elevating their work to the level of original research. Such opportunities for creativity are to be treasured in an almost exclusively analytic engineering curriculum.

Sprinkled throughout the book are 15 examples coming from design projects originated either by students, clients, or other research workers in optimization. Since these examples are used to illustrate theoretical principles in the text, they have been cleaned up considerably to avoid distracting the reader with details. It is, however, good experience for a student to grapple with such details of modeling and simplification, so some examples of original student design project proposals, edited only minimally, are given next. In a few, numerical values of the design parameters estimated by the student are given, but in most they are not. Part of the worth of the formulation phase for the student is finding reasonable cost and performance figures from the literature, vendors, or by theoretical analysis. Neither are all the problems completely formulated. As in practice, essential restrictions may be overlooked at first, and an important function of the procedures of this book is to find out when something has been forgotten before too much effort has been expended in vain.

The reader may wish to use one of the projects as a test for the theory. Preferably the reader will already have an interesting design project whose technology is familiar and whose solution would be of personal interest. Even then, a glance at the types of problem yielding to the methods of this book should indicate the range of potential application, providing motivation to learn the theory.

After the student projects are listed problems from the literature that also seem likely candidates for the procedures. As each piece of theory is developed, problems to which it would seem particularly applicable are indicated in the problem section at the end of the chapter in which the theory appears. Also instructions will be given for applying the chapter's theory to the student's individual design project, in the form of exercises marked with an asterisk (*). Don't wait for the theory; start looking for your project *now*.

NOTES AND REFERENCES

Works mentioned in the text are listed bibliographically at the end of each chapter, together with historical comments and discussion of related topics.

For this first chapter the references on optimization are grouped separately from those on design. The lists are selective rather than exhaustive, intended merely to suggest to the designer what has been done so far in optimization and give the operations analyst access to the design literature. The Quantitative Design group includes textbooks for the senior design courses preparing one for the study of Optimal Design, as well as books formulating design problems for subsequent numerical optimization—good sources for student optimal design projects. The Related Design Topics list reminds us all that there is more to design that just number processing.

Quantitative Design

Avriel, M., M. J. Rijckaert, and D. J. Wilde, *Optimization and Design*, Prentice-Hall, Englewood Cliffs, N.J., 1973.

Fox, R. L., *Optimization Methods for Engineering Design*, Addison-Wesley, Reading, Mass., 1971.

Johnson, R. C., *Optimum Design of Mechanical Elements*, Wiley, New York, 1961.

Johnson, R. C., *Mechanical Design Synthesis with Optimization Applications*, Van Nostrand Reinhold, New York, 1971.

Lothers, J. E., *Design in Structural Steel*, 3rd ed., Prentice-Hall, Englewood Cliffs, N.J., 1972.

Middendorf, W. H., *Engineering Design*, Allyn and Bacon, Boston, 1969.

Miele, A., Ed., *Theory of Optimum Aerodynamic Shapes*, Academic, New York, 1965.

Peters, M., and K. D. Timmerhaus, *Plant Design and Economics for Chemical Engineers*, 2nd ed., McGraw-Hill, New York, 1968.

Rudd, D. F., G. J. Powers, and J. J. Siirola, *Process Synthesis*, Prentice-Hall, Englewood Cliffs, N.J., 1973.

Rudd, D. F., and C. C. Watson, *Strategy of Process Engineering*, Wiley, New York, 1968.

Shigley, J. E., *Mechanical Engineering Design*, 2nd ed., McGraw-Hill, New York, 1972.

Siddall, J. N., *Analytical Decision-Making in Engineering Design*, Prentice-Hall, Englewood Cliffs, N.J., 1972.

Spotts, M. F., *Design of Machine Elements*, 4th ed., Prentice-Hall, Englewood Cliffs, N.J., 1971.

Stark, R. M., and R. L. Nichols, *Mathematical Foundations for Design: Civil Engineering Systems*, McGraw-Hill, New York, 1972.

Stoecker, W. F., *Design of Thermal Systems*, McGraw-Hill, New York, 1971.

Zener, C., *Engineering Design by Geometric Programming*, Wiley-Interscience, New York, 1971.

Related Design Topics

Adams, J. L., *Intellectual Blockbusting*, Stanford University Press, Stanford, 1975.

Asimov, M., *Introduction to Design*, Prentice-Hall, Englewood Cliffs, N.J., 1962.

Beakley, G. C., and E. G. Chilton, *Design: Serving the Needs of Man*, Macmillan, New York, 1974.

Dixon, J. R., *Design Engineering*, McGraw-Hill, New York, 1966.

Fuchs, H. O., and R. F. Steidel, Jr., *Ten Cases in Engineering Design*, Longman, London, 1973.

McKim, R. E., *Experiences in Visual Thinking*, Brooks/Cole, Monterey, Calif., 1972.

Murrell, K. F. H., *Human Performance in Industry*, Reinhold, New York, 1965.

Prince, G., *The Practice of Creativity*, Collier, New York, 1970.

Woodson, T. T., *Introduction to Engineering Design*, McGraw-Hill, New York, 1966.

Optimization Theory

Avriel, M., *Nonlinear Programming: Analysis and Methods*, Prentice-Hall, Englewood Cliffs, N.J., 1976.

Beightler, C. S., and D. T. Phillips, *Applied Geometric Programming*, Wiley, New York, 1976.

Bellman, R. E., *Dynamic Programming*, Princeton University Press, 1957.

Dantzig, G. B., *Linear Programming and Extensions*, Princeton University Press, 1963.

Duffin, R. J., E. L. Peterson, and C. Zener, *Geometric Programming*, Wiley, New York, 1967.

Fiacco, A. V., and G. P. McCormick, *Nonlinear Programming, Sequential Unconstrained Minimization Techniques*, Wiley, New York, 1968.

Garfinkel, R. S., and G. L. Nemhauser, *Integer Programming*, Wiley, New York, 1972.

Hammer, P. L., and S. Rudeanu, *Boolean Methods in Operations Research and Related Areas*, Springer-Verlag, Berlin, 1970.

Lasdon, L. S., *Optimization Theory for Large Systems*, Macmillan, New York, 1970.

Luenberger, D. G., *Introduction to Linear and Nonlinear Programming*, Addison-Wesley, Reading, Mass., 1973.

Nemhauser, G. L., *Introduction to Dynamic Programming*, Wiley, New York, 1966.

Wilde, D. J., and C. S. Beightler, *Foundations of Optimization*, Prentice-Hall, Englewood Cliffs, N.J., 1967.

Wilde, D. J., *Optimum Seeking Methods*, Prentice-Hall, Englewood Cliffs, N.J., 1964.

OPTIMAL DESIGN TERM PROJECT

Most of your grade will be determined by your ability to apply the ideas of this course to a design project of your own choice. The initial formulation should have 5 to 10 variables and be nonlinear. You can make up your own problem or else use one from the engineering literature. The easiest way to get a problem is to find a technical paper in which the objective function and constraints are already derived, but must be solved numerically. Your task is to see how much of it can be solved *analytically*, discovering simplifications and design rules. What cannot be solved analytically will be solved numerically. Have an initial brief proposal ready for me to see (date). Don't worry if you seem to have too many or too few variables; I'll modify the problem if necessary to make it acceptable. Describe the problem, together with the objective and all constraints, including graphical relations. Keep a copy of everything you turn in, because I'll retain your original documents so that I can refer to them as your project progresses. Don't be afraid to ask me for help, especially in the initial formulation. I enjoy consulting on these jobs, and if I don't think you need help, I'll tell you.

EXERCISES

1-1. *Optimization of a solar heated house* (D. Cautley and M. Kast, Stanford, 1974). Solar energy as a major energy source, now that low-cost fossil fuels are becoming scarce, will perhaps find its first large-scale utilization in space heating.

The relatively high capital cost of a solar energy system capable of providing the heating needs of a home normally met by natural gas or bunker grade fuel oil requires careful analysis of cost distribution within the system.

There are presently approximately 10 major variations in solar house design. One has been in daily operation for 10 years and if not optimum, it is the most workable. Our problem will consist of a simplified description of the entire house, its environment, and the solar radiation.

The inherent difficulty in harnessing solar energy is that you can only collect it when the sun is up, which may or may not coincide with your energy demand schedule. Therefore, both energy collection and storage systems are required, along with the necessary interfaces and controls.

The system studied collects solar radiation by heating a near-blackbody sheet metal wafer through which water is pumped. The water functions both as the heat-transfer medium and heat storage, being returned to a large storage tank at an elevated temperature. The water releases the heat to the rooms of the house as insulated louvered panels are swung away from the tank surface through convection and radiation (Fig. 1-3).

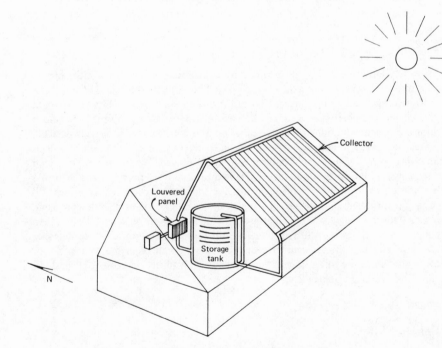

Figure 1-3 Solar house schematic

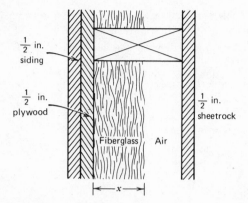

Figure 1-4(a)   Wall section

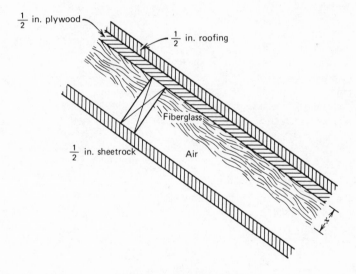

Figure 1-4(b)   Roof section

The size of the energy collection and storage systems must be matched to each other and to the heating requirements of the building. The heating requirement is a function of the house design, wall and roof construction, insulation thickness, window area, crack length around windows and doors, and of course weather conditions.

For this problem we have taken a house of reasonable dimensions and standard construction (stud wall) (see Fig. 1-4), imposed a "design day" environment on it, and left as variables the wall insulation thickness, roof insulation thickness, collector area (assuming 50% efficiency, closely approximating true values), length of weatherstripping used, storage tank volume, pumping and piping capacities and controls as a function of the others.

The objective function and constraints considered valid were developed as follows:

Building dimensions (consult sketch)

length	60 ft
width	30 ft
walls (height)	8 ft
south roof, width	25 ft
north roof, width	15 ft
3 doors, 7×3.5 ft	
6 windows 6×3 ft	

Design day
68°F mean inside temperature (2% vertical gradient)
35°F outside temperature
10 mph winds
outside convective coefficient $h_o = 6$ Btu/hr ft$^{2\circ}$ F
inside convective coefficient $h_i = 1$ Btu/hr ft^2 °F

Insulation
stud walls with $\frac{1}{2}$ in. sheetrock interior
space available for fiberglass insulation
 between walls = 3.5 in.
 between ceiling joists = 5.5 in.

$$\frac{\text{heat loss}}{\text{walls}} = \frac{T(\text{in}) - T(\text{out})}{RA}$$

$$= \frac{33}{4.5x} \text{ Btu/hr ft}^2$$

$$\frac{\text{heat loss}}{\text{roof}} = \frac{T(\text{in}) - T(\text{out})}{RA}$$

$$= \frac{39}{4.5x} \text{ Btu/hr ft}^2$$

Optimization Program Data
optimize solar heated house for cost

wall area	2730 ft^2	
roof area	2400	(available for collector = 1500 ft^2)
floor area	1800	
window area	108	
window crack length	108 ft	
door crack length	63 ft	

$$\text{heat loss, walls} = \frac{33}{4.5x} \text{ Btu/hr ft}^2 \quad (x = \text{thickness of fiberglass, in.})$$

$$\text{heat loss, roof} = \frac{39}{4.5x} \text{ Btu/hr ft}^2$$

Variables
x_1 = collector area, ft^2
x_2 = radius of storage tank, ft
x_3 = height of storage tank, ft
x_4 = wall insulation thickness, in.
x_5 = roof insulation thickness, in.
x_6 = weatherstripping seal length, ft

Cost ($)

$3.5x_1 + 150$
(1)

$6.28x_2^2$	$+ 15.7x_2x_3$	$+ 100$
(4) materials, tank top & bottom	(5) materials, sides	
$0.0045x_1^{3/2}$	$+ 100$	
(2) differential pump capacity cost		
$x_1^{1/2}$	$+ 31.4x_2^{1/2}x_3^{1/2}$	$+ 100$
(3) electronic controls, collector	(6) controls, storage	
$+ 137x_4$	$+ 72x_5$	$+ 0.2x_6$
(7) wall insulation	(8) roof insulation	(9) crack seal

Total (collector, storage, pumping, controls, insulation)

$$P_0 = \text{cost} - \$395 \quad (\text{constant value left out})$$

$$= \underset{(1)}{3.5x_1} + \underset{(2)}{0.0045x_1^{3/2}} + \underset{(3)}{x_1^{1/2}} + \underset{(4)}{6.28x_2^2} + \underset{(5)}{15.7x_2x_3}$$

$$+ \underset{(6)}{31.4x_2^{1/2}x_3^{1/2}} + \underset{(7)}{137x_4} + \underset{(8)}{72x_5} + \underset{(9)}{0.2x_6}$$

Constraints	Physical Interpretation
(1) $x_1 \leqslant 1500$	collector size limited by southern roof area
(2) $x_4 \leqslant 3.5$	wall insulation limited by stud width
(3) $x_5 \leqslant 5.5$	roof insulation limited by joist width
(4) $2x_2 x_3 \geqslant 150$	wall area of storage tank $\geqslant$ minimum bound for free convection
(5) $\begin{aligned} 770x_1 \geqslant & 48(2730(33/4.5x_4) + 2400(39/4.5x_5) \\ & + 1800(1.5) + 108(3.3/1.19) \\ & - (10^2 + 63)(0.075)(20)(0.24)(33)x_6) \end{aligned}$	heat input into water with collector efficiency $= 50\%$ two days of heating for house
(6) $770x_1 \leqq 1.2(140 - 68)(0.988)(\pi)x_2^2 x_3$	heat input per day $= 1.2 \times$ storage capacity (accounting for internal losses) inequality direction determined by requiring tank to store at least the amount of heat collected
(7) $x_6 \leq 171$	weatherstripping length limited by crack length

1-2. *Design of a proving ring* (N. N. A. Viet, Stanford, 1974). This project deals with the design of a proving ring for minimum manufacturing cost. A proving ring is a ring transducer that employs a sensitive micrometer for deflection measurement, as shown in Fig. 1-5. To obtain a precise measurement, one edge of the micrometer is mounted on a vibrating reed device R, which is plucked to obtain a vibratory motion. The micrometer is then moved forward until a noticeable damping of the vibration is observed. Deflection measurement may be made within ± 0.00002 in. with this method. The proving ring is widely used as a calibration standard for large tensile-testing machines.

Objective function

The objective function is the equation for the total cost of manufacturing the proving ring. The total cost C_t can be resolved into three main components.

The first is the fixed cost C_f, which includes charges for the micrometer and such services as handling and paper work. The fixed cost is an invariant component as far as optimum design is concerned, that is, it will be the same regardless of the final design.

The second component is the cost of material C_{mat}. Let c_1 be the cost of the material in dollars per pound. Thus $C_{mat} = c_1 w V$, where V is the volume of the ring. The value of the parameter c_1 will vary from material to material and possibly even with the shape of the stock available.

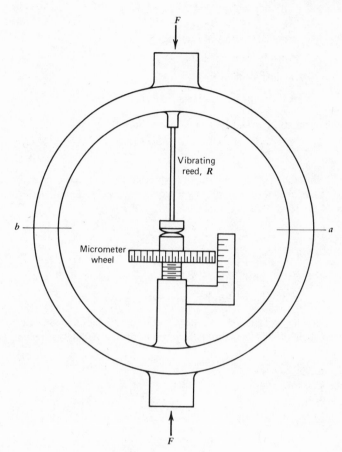

Figure 1-5 Proving ring

The third component of C_t is the cost of machining C_m. The process of making a proving ring consists of casting a tube (by investment casting or by die casting), then cutting the tube into thin rings, and finally machining each ring to acceptable tolerances. The cost of manufacturing the ring is actually a difficult part to account for precisely; it is primarily a function of material machining characteristics and the amount of machining. It would be reasonable to set the machining cost proportional to the total area of the ring, since the expenses of processes such as casting, cutting, or finishing vary directly with the area of the ring. $C_m = c_2 \cdot$ Area, where c_2 is the proportionality constant.

The equation of the total cost C_t is:

$$C_t = C_f + C_{\text{mat}} + C_m = C_f + c_1 wV + c_2 A$$

Let

$$r_o = \text{outside radius of the ring}$$
$$r_i = \text{inside radius of the ring}$$
$$l = \text{width of the ring}$$
$$C_t = C_f + c_1 wl\pi\left(r_o^2 - r_i^2\right) + c_2\left[2\pi\left(r_o^2 - r_i^2\right) + 2\pi l\left(r_o - r_i\right)\right]$$

The objective function would be to minimize the total cost.

$$p_0 = \min\left\{ c_1 wl\pi\left(r_o^2 - r_i^2\right) + 2\pi c_2\left[\left(r_o^2 - r_i^2\right) + l\left(r_o + r_i\right)\right]\right\}$$

To make the above objective function suitable for geometric programming, we can transform p_0 into a posynomial by introducing a new variable t, defined as the thickness of the ring. Thus

$$r_o - r_i = t$$
$$r_o^2 - r_i^2 = t^2 + ir_i t$$
$$p_0 \equiv \min\left\{ c_1 wl\pi\left(t^2 + 2r_i t\right) + 2\pi c_2\left[t^2 + 2r_i t + l\left(2r_i + t\right)\right]\right\}$$
$$p_0 \equiv \min\left\{ a_1 lt^2 + a_2 lr_i t + a_3 t^2 + a_4 r_i t + a_5 lr_i t + a_6 lt\right\}$$

$a_1,\ldots,a_6$ are the cost constants, which depend on factors peculiar to specific companies, such as overhead, types and quality of available machinery, and equipment.

Load constraint
The expression for the force applied to a proving ring is

$$F = \frac{64 EIy}{\pi d^3}$$

where

$$d = \text{outside ring diameter}$$
$$I = \text{moment of inertia about the centroidal axis of the ring section}$$
$$y = \text{deflection of the ring}$$
$$E = \text{Young's modulus of elasticity}$$

I can be computed as follows:

$$I = \pi l\left(r_o^4 - r_i^4\right) = \frac{\pi l\left(2r_i^3 t + 6r_i^2 t^2 + 4r_i t^3 + t^4\right)}{2}$$
$$d^3 = 8\left(r_i + t\right)^3 = 8\left(r_i^3 + 3r_i^2 t + 3r_i t^2 + t^3\right)$$

Thus

$$F = \frac{\pi Eyl\left(2r_i^3 t + 6r_i^2 t^2 + 4r_i t^3 + t^4\right)}{\left(\dfrac{\pi}{2} - \dfrac{4}{\pi}\right)\left(r_i^3 + 3r_i^2 t + 3r_i t^2 + t^3\right)}$$

The range of applied load for a commercialized proving ring is from 100 lb to 300,000 lb. However, we must add a safety factor to F: F should be at least equal to 300,000 lb.

The first constraint is

$$F \geq 300{,}000 \text{ (lb)}$$

Stress constraint

The maximum tensile stress occurs at points a and b (Fig. 1-5). The expression for tensile stress is

$$\sigma_{\text{tensile}} = \frac{2Ft^2}{lr}$$

The tensile stress of any material is subjected to an upper limit. Thus, our second constraint is

$$\sigma_{\text{tensile}} \leq \sigma_{\text{max}}$$

where σ_{max} depends on the material used to make the ring.

Accuracy constraint

Since the proving ring is used to calibrate large tensile-testing machines, it must have high accuracy. The uncertainty of the load ω_F/F must be less than or equal to a tolerance limit T_{max}, set up by the specific company which orders the proving ring.

$$\frac{\omega_F}{F} \leq T_{\text{max}}$$

By inspection of Eq. (1), and assuming that ω_l/l is insignificant compared to ω_F/F

$$\left(\frac{\omega_F}{F}\right)^2 = g\left(\frac{\omega_d}{d}\right)^2 + \left(\frac{\omega_l}{l}\right)^2 + \left(\frac{\omega_y}{y}\right)^2$$

which implies

$$g\left(\frac{\omega_d}{d}\right)^2 + \left(\frac{\omega_l}{l}\right)^2 + \left(\frac{\omega_y}{y}\right)^2 \leq T_{\text{max}}^2$$

Dimension constraints

The condition of making the ring a portable and handy instrument restricts the diameter of the ring to within certain dimensions. The diameter of the ring must be bounded by limits specified by the consumers.

$$d_{\text{min}} \leq d \leq d_{\text{max}}$$

Also, because of the casting process, the ring cannot be too thin ($t \geq t_{min}$) so that the molten metal can flow easily into the die, thus minimizing the presence of air bubbles and impurities in the metal. On the other hand, the ring must not be too thick ($t \leq t_{max}$) because, beyond a certain thickness, the formation of dendrites upon solidification can alter the yielding point as well as the hardness of the ring.

Deflection constraint

Since the micrometer used to measure the deflection of the ring has limited range, the ring should not be allowed to deflect beyond measurable ranges of the micrometer. Thus, the deflection y is subjected to an upper limit y_{max}, which depends on the characteristics of the micrometer or of any reading device.

1-3. *Chemical reactor fuel conservation* (P. Ducloux, Stanford, 1975). The problem is to minimize the quantity of heat to a chemical reactor. The gas phase reaction is $A \rightarrow B$, and the kinetics and thermodynamic properties are known (enthalpy ΔH, activation energy ΔE, and kinetic constant k_0). The reaction is isothermal at temperature T. If the reaction is endothermic, a heater is needed. If the reaction is exothermic, a cooler may be necessary if heat losses are not sufficient. The reactor feed is A at temperature T. Thus we have to heat A from room temperature to T before introducing it into the reactor.

There are three terms in the objective function:

(1) The heat losses by convection from the reactor, assumed proportional to time t and temperature T of the reaction.

$$L = \alpha T t$$

(2) The heat necessary to heat the feed to reaction temperature.

$$H_f = C_p (T - T_0) x_0$$

where C_p is the heat capacity per mole at constant pressure and x_0 is the number of moles introduced.

(3) The heat generated by the reaction.

$$H_r = y \Delta H,$$

where y is the number of moles of B, and H_r is positive (negative) if the reaction is endo (exo) thermic.

$$y = x_0 \left\{ 1 - \exp[-k_0 t \exp(-\Delta E / RT)] \right\}$$

The objective is therefore

$$z = \alpha T t + x_0 \Delta H \left\{ 1 - \exp[-k_0 t \exp(-E / RT)] \right\} + C_p (T - T_0) x_0$$

The parameters are $\alpha = 1$ cal/°K sec, $\Delta H = \pm 10^4$ cal, $T_0 = 300$°K, $\Delta E = 10^4$ cal, $R = 2$ cal/°K, $k_0 = 2(10^{-3})$ sec^{-1}, $C_p = 3$ cal/mole°K.

There are three constraints:

(1) The reactor temperature is bounded above by $T_m = 1000$°K.
(2) We want at least 100 moles of B ($y \geqslant 100$).
(3) The reactor pressure $p = x_0 RT/V$ is bounded above by $P_m = 50$ atm, where $V = 1$ m^3.

1-4. *Design of an annular power transmission cable* (K. Y. Wang, Stanford, 1974). The optimal design of high-voltage transmission lines of an electrical distribution network is outlined. Evidently the designer should reduce the number of tower structures for a given distance between hydroelectric station and consumer center. One way to achieve longer span distance is to use thicker wires. However, rising material cost forces use of as little conductor material as possible. Moreover, most of the alternating current flows near the outer surface. Hence a central core can be removed without seriously reducing the current carrying capacity. Thus the design problem involves two main considerations: the conductor cross section and the tower spacing.

The cross-sectional area of a conductor, assuming a linear current density $J = kr$, is $I = 2\pi k(r_o^3 - r_i^3)/3$. The physical relations of the shape of the cable between towers are, assuming a parabolic approximation to the theoretical caternary function, $d = wS^2/8t$; $l = S(1 + 8d^2/3S^2)$; $T = t + wd$; where w is the weight per unit length, and the other quantities are as shown in the Fig. 1-6.

Let: n = number of spans (an integer)
D = total distance
S = distance per span
m = number of cables strung on tower (typically 6)
C_1 = cost per pound of material
C_2 = cost per tower

Then the cost to be minimized is $C_1 m w Dl/S + C_2(n+1)$.

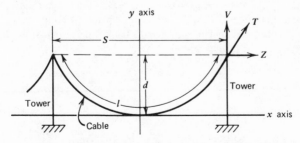

Figure 1-6 Power transmission cable and towers

There are seven constraints:

(1) The sag d cannot exceed 10 ft.
(2) The maximum stress is 11,000 psi for the aluminum alloy used.
(3) The maximum current density at the surface is 2400 amp/in.^{2}.
(4) The minimum current transmitted per cable is 1000 amp.
(5) The maximum outside diameter of the cable is 1.5 in.
(6) The minimum cable wall thickness is 0.2 in.
(7) The ultimate strength of the insulators is 15,000 lb.

1-5. *Optimal design of a reflected core nuclear reactor* (P. Braga, Pontifícia Universidade Católica do Rio de Janeiro, 1977). Lam and Townsend sought to maximize reflector economy and minimize the core material in a spherical core nuclear reactor; their paper gives the complete model development. There are four design variables, two equality constraints, one involving an inverse trigonometric function, and 13 inequality constraints. The equalities are energy balances; one inequality is connected with the probability that a fast neutron will reach thermal energies without capture; two are criticality considerations; two involve thermal diffusion length; one is the maximum flux constraint; the rest are practical and physical limitations.

These authors used a direct search algorithm and give designs. The purpose of this analysis is to see if the solutions found are global optima or not. None of the inequality constraints are active for the designs given, which suggests that they may be local, but not global, optima. See Lam, F. Y., and M. A. Townsend, "Optimal Design of a Reflected Core Nuclear Reactor," *Trans. ASME, J. Eng. Ind*, 995 (1975).

1-6. *Hydrodynamic journal bearings* (L. Heising, Stanford, 1975). Most designers use empirical guides rather than optimization in selecting main bearing dimensions because, although the dynamic analysis is well established, the relations between the various parameters are extremely complicated. Drawing on the work of Seireg and Ezzat, we can model the bearing to minimize the oil supply and the oil temperature rise with relative merit five to one.

We wish to choose the oil viscosity μ (Reynolds), the radial clearance c (in.), the bearing length l (in.), and the bearing average pressure p (psi), given the bearing load W (250 lb), the rotation speed N (rps), the half rotor mass M (lbf·sec^2/in.), and the journal diameter D (in.). Other variables are oil temperature rise Δt (°F), oil flow q (in.^{3}/sec), Sommerfeld number $S = \mu N/pc^2$, and length/diameter ratio $x = l/D$. The objective to be minimized is

$$y = \tfrac{1}{2} x^{-0.374} p S^{0.695 x^{-0.139}} + 19.5(10^3) c x^{0.952} S^{-0.1 x^{0.47}}$$

There are five sets of constraints.

(1) Film thickness $1.585 x^{0.913} S^{0.655 x^{0.0922}} \geqslant 5(10^{-5})$ in.
(2) Pressure $3(10^4)$ psi $\geqslant (0.76)^{-1} p x^{-0.62} S^{-0.24}$

(3) Viscosity $\mu \geqslant 10^{-7}$ Reynolds, which implies $p \geqslant 2.5(10^{-5})c^{-2}S^{-1}$
(4) Bearing length $0.25 \leqslant x \leqslant 0.50$ in.
(5) Stability $S^{-0.55}x^{-1.1} \geqslant 1564c$

It should be noted that in the original formulation the radius appears to vary, which is of course inconsistent with treating the diameter as a constant parameter. Seireg, A., and Ezzat, "Optimal Design of Hydrodynamic Journal Bearings," *Trans. ASME, J. Lubr. Tech.*, F **91**, 3, 1516 (July, 1969).

1-7. *A helical spring for a cam-driven system* (L. Mancini, Stanford, 1974). Design variables: $f =$ maximum spring force (lb), $m =$ minimum spring force (lb), $n =$ number of active coils, $d =$ wire diameter (in.), $c =$ coil diameter (in.).

$$\min y: y \equiv (f-m)c^{0.86}d^{-2.86}$$

subject to

$6.05(10^{-5})fc^{0.86}d^{-2.715} \leqslant 1$	yielding
$d \geqslant 3.55(10^{-2})c^2n$	surging
$m/f \leqslant 1 - 0.1256nc^2d$	buckling
$0.1 \leqslant m \leqslant 20$	minimum spring force
$0.05 \leqslant d/c \leqslant 0.25$	diameter ratio
$c + d \leqslant 3/2$	outside diameter
$c \geqslant d + 0.75$	inside diameter
$1.12n + 1.6 \leqslant d^{-1}$	spring pocket length
$n \geqslant 3$	minimum coils

The number of coils must be an integer, and the wire must have a standard diameter. It is interesting to compare the solution to this problem with that given by R. C. Johnson in *Mechanical Design Synthesis with Optimization Applications*, Van Nostrand Reinhold, New York, 1971. This student project evolved into L. J. Mancini and R. J. Piziali, "Optimal Design of Helical Springs by Geometrical Programming," *Engng. Optzn.* **2**, 1 (1976) 83–95.

1-8. *A viscosity pump* (E. Kerschen, Stanford, 1974). A viscosity pump functions on the principle that an applied shear stress causes motion in a viscous fluid (see Fig. 1-7). A two-dimensional solution of the Navier-Stokes equations leads to the following equations, in which the variables are: $Q =$ flow (ft³/sec), $\mu =$ viscosity (lbf-sec/ft²), $\omega =$ drum speed (rad/sec), $r =$ drum radius, $h =$ width of annulus (ft), $L =$ drum length (ft), $l =$ length over which fluid is in contact with rotating drum $[l = (2\pi - \beta)r]$.

Pressure increase $\Delta P = 12l\mu h^{-3}(\frac{1}{2}\omega rh - QL^{-1})$
Torque input $Llr(\frac{1}{2}h\Delta Pl^{-1} + \omega R\mu h^{-1})$
Pump efficiency $(6Q/L\omega rh)(1 - 2Q/L\omega rh)(4 - 6Q/L\omega rh)^{-1}$

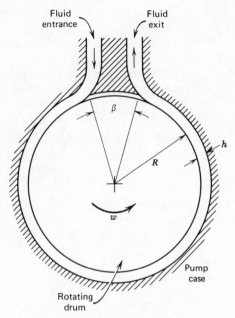

Figure 1-7 Viscosity pump

The problem is to design the pump giving minimum total cost for continuous operation for 12 h once a month with a specified flow rate and pressure rise in an extremely corrosive environment for a usable lifetime of 2 years. To construct a total cost function the following assumptions will be made: cost is proportional to (1) r^2; (2) $L^{3/2}$; (3) torque^2; (4) ω^3; (5) pump life is 2 years with no salvage value; (6) annual interest rate is 6%. Assuming simple interest and using the end of the first year as a value base, we must multiply the initial pump cost by 1.08. The cost of operating for two years will be calculated under the assumptions of \$0.02/kwh power cost and 85% drive motor efficiency. Then the operating cost to be minimized is $3.24(10^{10})\Delta P\omega hrL + 6.49(10^{10})\omega^2 r^2\mu lLh^{-1}$ in dollars.

Sobersky, Acosta, and Hauptman, *Fluid Flow*, Problem 7.2.

CHAPTER 2

Specifications
and Constraints

A man whose arm is on fire cannot think of
meditation for even a second.

Geshe Rabten, quoted by Ngo-Dhro in

Preliminary Practices of Tibetan Buddhism

In simple situations, good engineers generate optimal designs "intuitively," as they would say. This mental process is not really intuition, but rather an optimization, unconscious because mathematically trivial, of an objective too obvious to bother stating quantitatively, or even verbally. This chapter analyzes one aspect of "designer's intuition," placing it on a firm but simple theoretical footing so that engineers can use this power to its fullest, knowing how to keep it rigorously sound.

Consider an example where designer's intuition quickly generates an optimal design. Suppose the diameter of a vertical steel wire is to be selected to support a tensile load of 1000 kg. Knowing the tensile strength of the wire to be 2000 kg/cm^2, the designer simply divides it into the load to find a cross sectional area of 0.5 cm and "intuitively" chooses the smallest standard size wire having at least this area. A smaller wire would break; a larger one would waste material. In reasoning this way, the designer has, perhaps unconsciously, taken wire weight as the objective to be minimized. As a mathematical problem, this optimization is trivial because both weight and load capacity of the wire increase with diameter. Clearly then, the lightest wire is the smallest available that can carry the load.

If there is anything worthy of being called "engineering design" in this problem, it is the estimation of the load and the knowledge of the wire strength, together perhaps with the calculation of the minimum area, not the completely trivial "optimization." Yet the problem illustrates a powerful property occurring widely in design engineering, that of *monotonicity*. Both weight and load capacity, because they increase strictly with wire diameter, are examples of *monotonically increasing* functions. The implicit requirement that the load capacity be no less than the 1000-kg load is an example of what an engineer would call a *specification* and a mathematician a *constraint*. In this case the specification is known in detail, in that it can be written as a precise mathematical inequality involving diameter d:

$$(2000 \text{ kg/cm}^2)\left(\frac{\pi}{4}\right)(d \text{ cm})^2 \geq 1000 \text{ kg}$$

On the other hand, the weight is not given as a detailed function of diameter; all that is known is its monotonicity. This, however, is enough to establish that the wire diameter must be as small as possible without violating the constraint. The precision of the specification is of course needed to determine an exact lower bound on d by solving the constraint inequality.

$$d = \left[\frac{1000(4/\pi)}{2000}\right]^{1/2} = 0.79 \text{ cm}$$

But more detail on the weight dependence is unnecessary.

This powerful but simple concept of monotonicity generalizes easily to problems with more than one variable. As shown in the next section, the way in which several variables are handled one at a time, known as *partial optimization*, can be used even in the absence of monotonicity. But when monotonicity is present, partial optimization can rapidly simplify or solve design problems looking quite difficult at first glance.

2.1.　PARTIAL OPTIMIZATION

Often an apparently complicated design problem can be simplified considerably by optimizing the design variables one at a time, a procedure known as *partial optimization*. Consider, for example, a pressure vessel design problem formulated by Unklesbay, Staats, and Creighton as a four variable constrained problem to be optimized by a special technique known as geometric programming. Partial optimization will instead solve the problem using only the simplest of algebraic manipulations.

A compressed air storage tank with a working pressure of 300 lb/in.[2] (20.4 atm) and a volume of 750 ft[3] (21.2 m[3]) is to be designed. The vessel is to be a vertical cylindrical shell with a head at each end. The shell is to be fabricated from two rolled plates welded longitudinally to form a cylinder. The forged heads are to be welded to the shell. All material is to be carbon steel, SA-212 Grade B. The problem is to be solved for both flat and for hemispherical heads. All welds are to be fully radiographed single-welded butt joints with a backing strip. Space limits the shell length to no more than 20 ft (610 cm). The vessel is to conform to the American Society of Mechanical Engineers (ASME) Boiler and Pressure Vessel Code. Figure 2-1 shows the vessel with nomenclature.

First consider the influence of the head thickness h when the other three design variables s (shell thickness), r (inside radius of vessel), and l (shell length) are held constant. The total cost $c(h,s,r,l)$, which is to be minimized, has only one term, the head cost $c(h,r)$, depending on h. Hence the total is minimized if and only if this head cost is minimum with respect to

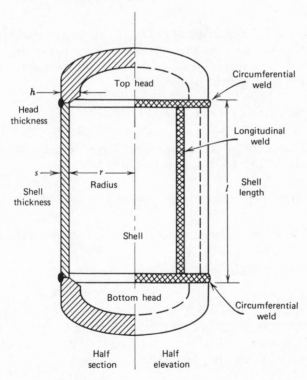

Figure 2-1 Vertical pressure vessel with dished heads

(henceforth abbreviated "wrt") h, no matter what values the other variables may assume.

Formally, let the greatest lower bound on $c(h,r)$, with r regarded as fixed, be written $\min_h c(h,r)$, so that

$$c(h,r) \geqslant \min_h c(h,r)$$

The quantity $\min_h c(h,r)$ is called the *partial minimum* of $c(h,r)$ wrt h, and the process of finding $\min_h c(h,r)$ is called *partial minimization* wrt h. Notice that $\min_h c(h,r)$ is a function of r, although not h, the latter being said to have been *minimized out* of $c(h,r)$. Now consider the total cost, consisting of head cost $c(h,r)$ plus the cost of all other items, written $c(s,r,l)$ because by definition they do not depend on h. Partial minimization of this total cost wrt h gives by definition

$$\min_h \left[c(h,r) + c(s,r,l) \right] = \min_h c(h,r) + \min_h c(s,r,l)$$

$$= \min_h c(h,r) + c(s,r,l)$$

because $c(s,r,l)$ is independent of h. This partial minimum of the total cost depends on all variables except h, the one that was minimized out.

Ultimately, the designer wants to minimize the total cost wrt all variables, the result being written

$$\min_{h,s,r,l} \left[c(h,r) + c(s,r,l) \right]$$

This can be obtained by taking the partial minimum wrt h and minimizing it wrt the other variables, a process that may be written

$$\min_{s,r,l} \left\{ \min_h \left[c(h,r) + c(s,r,l) \right] \right\} = \min_{s,r,l} \left[\min_h c(h,r) + c(s,r,l) \right]$$

That the two minima must be equal is expressed by the equation

$$\min_{h,s,r,l} \left[c(h,r) + c(s,r,l) \right] = \min_{s,r,l} \left[\min_h c(h,r) + c(s,r,l) \right]$$

The right member indicates that the original problem in four variables is to be solved first as a partial optimization problem in h. This will be seen to have two advantages: first, only head cost need be considered, and, second, h has only one very simple constraint, namely

$$h \geqslant Hr$$

where H is a parameter depending on the known design parameters working pressure P, allowable stress S, joint efficiency E, and head attachment factor C. Specifically, in this problem

$$H \equiv \begin{cases} 2\left(\dfrac{CP}{S}\right)^{1/2} = \dfrac{2(0.25)(300)}{17500^{1/2}} = 130(10^{-3}) \\ \quad \text{for flat heads} \\ P(2S - 0.2P)^{-1} = 300\left[2(17500) - 0.2(300)\right]^{-1} = 8.6(10^{-3}) \\ \quad \text{for hemispherical heads} \end{cases}$$

Henceforth, given design *parameters* will be symbolized by capital letters to distinguish them from the design *variables*, indicated by lower-case letters. Parameters are given in advance; variables are to be determined by the designer.

The resulting partial optimization problem is

$$\min_{h} c(h,r)$$

subject to

$$h \geqslant Hr$$

where r is regarded as a temporarily fixed parameter, and H is treated as a parameter rather than a constant so that the results may be generalized to different values of the design parameters. Now the head cost will certainly increase wrt h, being in fact

$$c(h,r) = \left(C_h r^2\right)h$$

where C_h is a known positive parameter (3.554 in this specific problem). Hence

$$\min_{h} c(h,r) = C_h r^2 \left(\min_{h} h\right)$$

Therefore for minimum cost, h must be as small as possible, that is, at its greatest lower bound Hr, which is proportional to the as yet unknown radius r. The subscript $\star$ indicates that h is at its minimizing value (read "h substar")

$$h_{\star} = Hr$$

The minimized head cost is therefore

$$\min_{h} c(h,r) = c(h_\star,r) = c(Hr,r) = C_h r^2 (Hr) = (C_h H) r^3$$

The parentheses emphasize that the product $C_h H$ is a parameter, not a design variable.

Next optimize partially wrt the shell thickness s. Other things being equal, or more precisely, with h, r, and l held constant, both shell cost and welding cost increase wrt s. Hence s should be as small as possible. Now the ASME Boiler and Pressure Vessel Code imposes only one constraint on shell thickness, namely

$$s \geqslant K_s r$$

where K_s is an parameter depending on P, E, and S. Specifically, in this problem

$$K_s = \frac{P}{(2SE - 0.6P)} = \frac{300}{[2(17500)(0.9) - 0.6(300)]} = 9.59(10^{-3})$$

Therefore the cost is minimized when s is at this greatest lower bound

$$s_\star = K_s r$$

Thus

$$\min_{s,r,l} \left[c(s,r,l) \right] = \min_{r,l} \left\{ \min_{s} \left[c(s,r,l) \right] \right\} = \min_{r,l} \left[c(s_\star, r, l) \right]$$

$$= \min_{r,l} \left[c(K_s r, r, l) \right]$$

A more specific result is obtained by breaking the cost $c(s,r,l)$, which is the total cost less the head cost, into its three components: shell material cost, radial weld cost, and longitudinal weld cost. Shell material cost is proportional to shell volume $2\pi rls$. Incorporation of 2π into the other parameters gives $C_s rls$, where C_s is a cost parameter that must be estimated. Since the width of a weld is proportional to the shell thickness s, as is the weld depth, the radial weld cost is $C_r rs^2$ and the longitudinal weld cost is $C_l ls^2$, where C_r and C_l are estimated cost parameters. Hence

$$c(s,r,l) = C_s rls + C_r rs^2 + C_l ls^2$$

Therefore the sum of these costs, minimized wrt s, is

$$c(s_\star, r, l) = c(K_s r, r, l) = (C_s K_s) r^2 l + (C_r K_s^2) r^3 + (C_l K_s^2) r^2 l$$

a function strictly increasing wrt both r and l, as is the head cost $(C_h H)r^3$.

The following functional equations summarize the partial optimizations so far. Notice how the constraints used are placed below the minimization operator.

$$\min_{h,s,r,l} \left[c(h,r) + c(s,r,l) \right]$$

$$= \min_{r,l} \left[\min_{h \geqslant Hr} c(h,r) + \min_{s \geqslant K_s r} c(s,r,l) \right]$$

$$= \min_{r,l} \left[c(Hr,r) + c(K_s r,r,l) \right]$$

$$= \min_{r,l} \left[(C_h H)r^3 + (C_s K_s)r^2 l + (C_r K_s^2)r^3 + (C_l K_s^2)r^2 l \right]$$

$$= \min_{r,l} \left[(C_h H + C_r K_s^2)r^3 + (C_s K_s + C_l K_s^2)r^2 l \right]$$

To complete the minimization wrt r and l, the remaining constraints must be examined.

The radius r is constrained by the geometry, since the vessel volume V is a given parameter

$$V = \begin{cases} \pi r^2 l & \text{flat heads} \\ \pi r^2 l + 4\pi r^3/3 & \text{hemispherical heads} \end{cases}$$

Both functions increase strictly wrt r and l, as does the partially minimized total cost. Since r appears in no other constraints, it will achieve its greatest lower bound at the minimum cost. Therefore it would make no difference if the strict equality were replaced by the inequality

$$V \geqslant \begin{cases} \pi r^2 l & \text{flat} \\ \pi r^2 l + 4\pi r^3/3 & \text{hemispherical} \end{cases}$$

This tells the designer that the *optimal* cost of the vessel strictly increases wrt the required volume V. In other words the minimum cost vessel will have the smallest volume needed to do the job. This is because the minimum cost design will always require strict equality.

The volume equation can be solved explicitly for l, although not for r, the solution being

$$l = \begin{cases} (V\pi^{-1})r^{-2} & \text{flat} \\ (V\pi^{-1})r^{-2} - \left(\dfrac{4}{3}\right)r & \text{hemispherical} \end{cases}$$

Elimination of l from the cost gives, for hemispherical heads,

$$c = V\pi^{-1}(C_s K_s + C_l K_s^2) + (C_h H + C_r K_s^2 - 4/3)r^3$$

For flat heads the negative term would be deleted. Notice that the constant term $V\pi^{-1}(C_s K_s + C_l K_s^2)$, which is the cost of the shell including longitudinal welding cost, is completely unaffected by the vessel radius in the optimum design. This is an example of a useful universal design rule produced by applying optimization theory to an engineering problem.

Since the coefficient $(C_h H + C_r K_s^2 - 4/3)$ of r^3, certainly positive for flat heads, is also positive for hemispherical heads, the cost strictly increases wrt r. Hence the tank should have the smallest diameter possible. In this problem, the radius is bounded below because of an upper bound L on the shell length.

$$l = \begin{cases} (V\pi^{-1})r^{-2} & \text{flat} \\ (V\pi^{-1})r^{-2} - \left(\dfrac{4}{3}\right)r & \text{hemispherical} \end{cases} \Bigg\} \leqslant L$$

To minimize cost then, the radius must be at its lower bound, which is equivalent to having the shell length as large as allowed.

The optimal design is therefore obtained sequentially from the following equations.

$$l_\star = L$$

For flat heads,

$$r_\star = \left(\frac{V}{\pi L}\right)^{1/2}$$

For hemispherical heads, $r_\star$ is the solution of the cubic equation

$$\left(\frac{4}{3}\right)r_\star^3 + Lr_\star^2 - \left(\frac{V}{\pi}\right) = 0$$

Since the left member strictly increases wrt $r_\star$ for $r_\star > 0$, there is exactly one solution. The optimal head and shell thicknesses are determined by

$$h_\star = Hr_\star$$
$$s_\star = K_s r_\star$$

These formulas collectively form a *design procedure*.

In the example for flat heads, in which the key parameters are $L = 240$ in. (610 cm), $V = 750$ ft^3 (21.2 m^3), $H = 130(10^{-3})$, and $K_s = 9.59(10^{-3})$, the optimal design is $l_\star = 240$ in. (610 cm), $r_\star = 41.5$ in. (105 cm), $h_\star = 5.43$ in. (138 mm), and $s_\star = 0.80$ in. (20 mm), verifying the design of Unklesbay, Staats, and Creighton obtained by the more complicated technique of geometric programming (Exercise 2-1). The hemispherical head case is left as Ex. 2-2. This case, only mildly more difficult because of the cubic equation to be solved, could not be solved directly by geometric programming for technical reasons, forcing Unklesbay, Staats, and Creighton to use a successive approximation technique.

Perceptive designers will object to this design because it calls for plate dimensions ($s = 0.80$ in., for example) which, being nonstandard, are either not available or very expensive. This valid objection will be dealt with in Sec. 2.4, in which an optimal design using standard sizes will be produced by modifying slightly the preceding design procedure.

Notice that this design is determined totally without knowing the values of the cost parameters. Thus the designer knows to make the vessel as long as space or fabricating equipment permits, with all other dimensions determined by the ASME code, every specification being treated as a strict equality. When, as in this example, there are at the optimum exactly as many strict equalities as there are variables, the design is said to be *constraint bound*. In this event, the objective does not need to be known in detail.

2.2. CODES, SPECIFICATIONS, AND CONSTRAINTS

For a design to be constraint bound is far from unusual. Circumstances that might bring this about are discussed in this section to show designers how to exploit the freedom of constraint-bound designs from the need for detailed economic information. The history of pressure vessels and their design illustrates how the informal interplay of economics and safety evolved into a rigorous code in which designs are determined entirely by the specifications rather than the economics.

The Age of Steam needed containers able to hold air, water, or vapor at relatively low pressure or vacuum. Since wood and ceramics are inadequate materials, metals such as bronze, brass, or cast iron plates were in the early days bolted or riveted together to form condensers, cylinders, and boilers. Operating engineers would increase the load on a steam engine cautiously until seams began leaking. As the equipment aged and corroded, performance was trimmed. By trial and error, engineers eventually learned to predict reasonable pressure limits for given vessel dimensions.

As engines became more powerful and compact, demanding higher steam pressures, equipment failures lead not just to leaks, but to spectacular explosions. The public imposed limitations on boiler pressures with guarantees of safety. Responsible engineers combined theoretical stress analysis with control of material strength and fabrication procedures into a precise code for pressure vessel construction and design. Economics also had an influence. A consumer, say a steamship company, would determine its needs as pressure, temperature, flow, volume, and power specifications. Then they would invite bids from vessel fabricators who naturally would try to avoid unnecessary expense. The obvious monotonicity of the cost with respect to the design variables, namely plate and head dimensions, forced fabricators to make everything just as small as they could get away with. The first simple codes prevented the most obvious modes of failure, and research into subsequent accidents lead to more constraints and safety factors. Addition of inspection and certification procedures has finally lead to the present situation where vessel failures are suitably rare.

This pattern of development from empirical experiment to detailed code has been repeated for many technological devices, especially the simple ones in widespread use. Economic monotonicity, where apparent, impels the designer to be sparing, until failure, accident, or poor performance shows that more expense is unavoidable. As consumers become more demanding, new modes of failure are encountered requiring additional restrictions. Quantitative research gradually supplants early qualitative experience until a detailed design code emerges. When failures do not affect public welfare, an official code may not be promulgated, the needed design information being instead published in the engineering literature. For relatively new devices in competitive industries, design constraints may remain unpublished, residing instead in secret company files.

Little wonder then that many designs are constraint bound (Exercise 2-3). Knowing this, the designer can concentrate on mastering monotonicity analysis before the more difficult aspects of optimization theory. The examples in this chapter provide practice in monotonicity analysis, even where dimensions are confined to a few standard sizes. The chapter concludes with a more general mathematical approach to monotonicity so that it can be recognized easily even in complicated functions.

2.3. MONOTONICITY ANALYSIS

Here the pressure vessel design problem is redone to include all types of head: dished, flat, and hemispherical. All cost, stress, and volume parameters for the vessel heads are given the same symbols, since they have different values for the various types of head.

The optimization will demonstrate a simple version of *monotonicity analysis* in which the only information used about the various functions is whether they are increasing or decreasing wrt the variable. For example, the head cost, instead of being written $C_h r^2 h$, is written $\varphi_h(r,h)$, which is read as "increasing wrt both r and h." The symbol φ will be reserved for increasing functions, with subscripts identifying the role of φ in the problem. In this notation, the hypothetical function $13x/y^2$ for $x,y>0$ would be written $\varphi(x,y^{-1})$, for a function decreasing in y is increasing in y^{-1}. If a variable appears with a circumflex ($\hat{\ }$) then the function is neither increasing nor decreasing wrt that variable. For example, $xy^{-2} - x^2 y$ is written $(\hat{x}, y^{-1})$ for $x,y>0$.

In this notation the cost of the pressure vessel can be written $\varphi_c(s,h,l,r)$, since cost increases wrt all four design variables. The ASME specification for the shell thickness is then

$$s \geqslant \varphi_s(r)$$

For head thickness it is

$$h \geqslant \varphi_h(r)$$

The volume relation is

$$\varphi_v(r,l) \geqslant V.$$

Finally, the total length constraint is

$$\varphi_l(l,r,h) \leqslant L.$$

Partial optimization wrt shell thickness immediately gives

$$s_\star = \varphi_s(r)$$

That is, the shell thickness is made as small as the ASME Code permits, with thickness increasing with vessel radius. The cost function is now $\varphi_c(\varphi_s(r),r,h,l)$ which still increases wrt r, h, and l, since an increasing function of an increasing function is certainly increasing also (proof left as an exercise).

Subsequent partial optimization wrt h would require h to attain a lower bound. This time there are two constraints on h, the code constraint being a lower bound; the total length constraint, an upper bound. Since cost increases wrt h, the lower bound must be active at the minimum

$$h_\star = \varphi_h(r)$$

The cost function is now $\varphi_c[\varphi_s(r),\varphi_h(r),r,l]$, which increases in both r and l. Both of the constraints remaining now involve these variables:

$$\varphi_v(r,l) \geqslant V$$

$$\varphi_l(l,r,\varphi_h(r)) \leqslant L$$

Partial optimization wrt r would force the volume constraint to be active, since the total length constraint bounds r above rather than below. For that matter, partial optimization wrt l would give the same result by the same reasoning. Hence the optimal values of r and l must satisfy

$$\varphi_v(r_\star,l_\star) = V$$

By the Implicit Function Theorem there exists a *decreasing* function $l = \varphi(r^{-1})$ of r such that

$$\varphi_v(r,l) = \varphi_v(r,\varphi(r^{-1})) = V$$

Thus the cost function can be written functionally in terms of r alone as $\varphi_c(\varphi(r^{-1}),r,\varphi_h(r))$, which, being composed of both increasing and decreasing functions of r, does not lend itself to further monotonicity analysis. At this point one must for the first time consider the specific functions involved. They are first, the cost

$$\varphi_c = c = C_h r^2 h + C_s rls + C_r rs^2 + C_l ls^2$$

respectively, the costs of the heads, the shell, the radial welds, and the longitudinal weld. Also the constraint functions are

$$\varphi_s(r) \equiv K_s r$$

$$\varphi_h(r) \equiv Hr$$

$$\varphi_v(r,l) \equiv \pi r^2 l + K_h r^3$$

$$\varphi_l(l,r,h) \equiv l + 2h + K_r r$$

where K_h and K_r are parameters for computing the volume and depth of the heads, respectively. These would be zero for flat heads.

The volume relation $\varphi_v(r,l)$ can be solved explicitly for l as a function of r.

$$l \equiv \varphi(r^{-1}) = (V\pi^{-1})r^{-2} - (K_h \pi^{-1})r$$

This is, as predicted by the Implicit Function Theorem, a decreasing function of r for positive r. Elimination of s, h, and l from the cost function gives

$$c = V\pi^{-1}(C_sK_s + C_lK_s^2) + (C_hH + C_rK_s^2 - K_h\pi^{-1})r^3$$

which increases wrt r provided

$$C_hH + C_rK_s^2 - K_h\pi^{-1} > 0$$

This condition is satisfied in practice for metal vessels.

Since the cost increases in r, it must be bounded below by the remaining constraint on length, which therefore must be active at the minimum. Hence the optimum value $r_\star$ of the radius must satisfy

$$(-K_h\pi^{-1} + 2H + K_r)r_\star^3 + Lr_\star^2 - V\pi^{-1} = 0$$

Since $K_h \leqslant 4\pi/3$, the parameter $(-K_h\pi^{-1} + 2H + K_r)$ is positive; this function strictly increases wrt $r_\star$ when $r_\star > 0$. Also the function is negative for $r_\star = 0$, so there is a unique root in the range of interest. Incidentally, the total length constraint, in terms of r, is

$$l + 2h + K_r r = V\pi^{-1}r^{-2} + (-K_h\pi^{-1} + 2H + K_r)r \leqslant L$$

The left member decreases in r, so the inequality provides a lower bound on r as required by the minimization of the objective, which increases in r. With $r_\star$ known, the other design variables are found by

$$l_\star = L - (2H + K_r)r_\star$$
$$h_\star = Hr_\star$$
$$s_\star = K_s r_\star$$

Again the design is constraint bound, and all the designer need know about the objective is its monotonicity. This is why pressure vessels have long been designed optimally without sophisticated optimization theory.

2.4. STANDARD SIZES

In the pressure vessel design problem, as in many realistic design situations, the design variables usually cannot assume the values computed by the preceding methods. This is because the physical elements of the vessel are restricted to standard sizes. Thus the vessel radius may have to be a multiple of 3 in., and the shell length may be determined by the width of a steel plate that comes only in multiples of 6 in. to a maximum of 20 ft. Shell and head thicknesses may also come in standard sizes with 0.25 in. increments.

When this happens, the constraints, written as inequalities, can be used to find the optimum *standard* sizes. In the pressure vessel example, let the solution to the optimum radius equation be written $\bar{r}$ (read r bar), since it will not be the optimal value $r_\star$ unless, by coincidence, all values of the design variables computed previously happen to coincide with the standard sizes available. In fact, the monotonicity of the cost wrt r guarantees that $r_\star \geqslant \bar{r}$.

Let $\lceil \bar{r} \rceil_1$ (read $\bar{r}$ rounded up sub one) represent the smallest standard vessel radius still exceeding $\bar{r}$. The symbol $\lceil \; \rceil$ indicates rounding up from the argument, in this case $\bar{r}$. Then the shell length is bounded below by the volume constraint, evaluated at $\lceil \bar{r} \rceil_1$.

$$l \geqslant (V\pi^{-1})\lceil \bar{r} \rceil_1^{-2} - (K_h\pi^{-1})\lceil \bar{r} \rceil_1$$

Let $\lceil \bar{l} \rceil_1$ be the smallest standard shell length that satisfies this inequality. With r fixed at $\lceil \bar{r} \rceil_1$, the length is bounded above by the total length constraint

$$l \leqslant L - K_r\lceil \bar{r} \rceil_1 - 2\lceil h \rceil_1$$

where $\lceil h \rceil_1$ is the smallest standard head thickness satisfying

$$h \geqslant H\lceil \bar{r} \rceil_1$$

Let $\lfloor \bar{l} \rfloor_1$ (read l rounded down sub one) be the largest standard shell length satisfying the length constraint, where in general $\lfloor \; \rfloor$ indicates rounding down. Then a standard length exists that will satisfy all constraints if and only if

$$\lceil \bar{l} \rceil_1 \leqslant \lfloor \bar{l} \rfloor_1$$

When there is a feasible standard size, the optimum is $r_\star = \lceil \bar{r} \rceil_1, l_\star = \lceil \bar{l} \rceil_1, h_\star = \lceil h \rceil_1, s_\star = \lceil s \rceil_1 \equiv \lceil K_s\lceil \bar{r} \rceil_1 \rceil$, where $\lceil s \rceil_1$ is the smallest shell thickness satisfying $s \geqslant K_s\lceil \bar{r} \rceil_1$.

There may not be a feasible standard shell length, however, which is what happens when $\lceil \bar{l} \rceil_1 > \lfloor \bar{l} \rfloor_1$. In this situation, the vessel radius must be rounded up again to the smallest standard size greater than $\lceil \bar{r} \rceil_1$. Thus define

$$\lceil r \rceil_2 \equiv \lceil \lceil \bar{r} \rceil_1 \rceil > \lceil \bar{r} \rceil_1$$

Then

$$\lceil h\rceil_2 \equiv \lceil H\lceil r\rceil_2 \rceil \geqslant H\lceil r\rceil_2$$

$$\lceil l\rceil_2 \equiv \lceil (V\pi^{-1})\lceil r\rceil_2^{-2} - (K_h\pi^{-1})\lceil r\rceil_2 \rceil$$

$$\lfloor l\rfloor_2 \equiv \lfloor L - K_r\lceil r\rceil_2 - 2\lceil h\rceil_2 \rfloor$$

and if $\lceil l\rceil_2 \leqslant \lfloor l\rfloor_2$, then $r_\star = \lceil r\rceil_2$, $l_\star = \lceil l\rceil_2$, $h_\star = \lceil h\rceil_2$, and $s_\star = \lceil K_s\lceil r\rceil_2 \rceil$. Otherwise, when $\lceil l\rceil_2 > \lfloor l\rfloor_2$, no feasible standard length exists for $\lceil r\rceil_2$ and r must be rounded up again, until eventually a feasible standard design is found on the next round in Fig. 2-2 (Exercise 2-5).

Despite the forbidding notation, this method is easy to apply and gives a good, although not necessarily optimal standard size design. The total length constraint, as shown in Fig. 2-2, forces the upper bound on shell length to be nonincreasing wrt radius. Any shell length infeasible for a given radius will thus remain infeasible for larger ones. Therefore, a good radius is the smallest for which a standard shell length gives the volume required without violating the length constraint. This simple approach will solve many, but not all, design problems involving standard sizes. More difficult problems of this type are treated in Chapter 8.

Notice that no economic information is needed to design the vessel in this constraint-bound case. The constraints determine the design completely, although the monotonicity of the objective function had to be

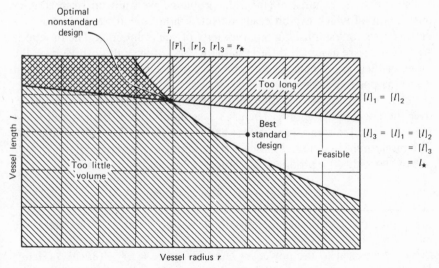

Figure 2-2 Standard vessel dimensions. Grid points represent standard sizes

assumed. In a given case the fabricator's designer would know what standard sizes are available. The user's designer should just specify the constraints and send the job out for bids.

2.5. THE MATHEMATICS OF MONOTONICITY

Monotonicity is an easily detected property of the functions in optimization problems that, when present, simplifies things greatly. A function $f(\mathbf{x})$ is said to be *increasing wrt* x_1, one of its independent variables, if and only if $\Delta f/\Delta x_1 > 0$ for all $\mathbf{x} > \mathbf{0}$ and for all $\Delta x_1 \neq 0$. This definition includes the case where f is differentiable wrt x_1, in which case $\partial f/\partial x_1 > 0$ for all $\mathbf{x} > \mathbf{0}$. If $f(\mathbf{x})$ increases wrt x_1, then $-f(\mathbf{x})$ is said to *decrease* wrt x_1. Notice that the same function can increase in one variable while decreasing in another. Finally, $f(\mathbf{x})$ is called *independent* of x_1 if and only if $\Delta f/\Delta x_1 = 0$ for all $\mathbf{x} > \mathbf{0}$ and all $\Delta x_1 \neq 0$.

A set of functions is said collectively to be *monotonic* wrt x_1 if and only if every one of them is either increasing, decreasing, or independent wrt x_1. The term *monotonic* is reserved for sets of functions that are not all independent. Similarly, a set of functions is said to be *increasing* (*decreasing*) wrt x_1 if and only if all functions are increasing (decreasing) wrt x_1. If one function is increasing and the other decreasing, both wrt x_1, they are said to have *opposite* monotonicity wrt x_1. Two functions that either both increase or both decrease are said to have the *same* monotonicity wrt x_1.

The functions in most engineering problems are built up from simpler ones, many of which exhibit easily detected monotonicity. Simple rules are derived now for establishing monotonicity of such functions. The functions studied here are assumed differentiable to first order and positive over the range of their arguments. Then f increases (decreases) wrt x_1 if and only if $\partial f/\partial x_1$ is positive (negative).

Let f_1 and f_2 be two positive differentiable functions monotonic wrt x_1 over the positive range of $\mathbf{x}$. Then $f_1 + f_2$ is monotonic if both f_1 and f_2 are monotonic in the same sense, a fact easily proven by direct differentiation. The monotonic functions f and $-f$ have opposite monotonicities.

Now consider the product $f_1 f_2$. Since

$$\partial (f_1 f_2)/\partial x_1 = f_1 (\partial f_2/\partial x_1) + f_2 (\partial f_1/\partial x_1)$$

the product $f_1 f_2$ will be monotonic if both f_1 and f_2 have the same monotonicities.

Let f be raised to the power α. Then $\partial (f^\alpha)/\partial x_1 = \alpha f^{\alpha-1}(\partial f/\partial x_1)$. Hence f^α will have the same monotonicity as f whenever α is positive. If $\alpha < 0$, f^α will have opposite monotonicity.

Finally consider the composite function $f_1(f_2)$. Differentiation by the chain rule gives $\partial f_1(f_2)/\partial x_1 = (\partial f_1/\partial f_2)(\partial f_2/\partial x_1)$. Hence the composition $f_1(f_2)$ is also monotonic. It increases (decreases) whenever f_1 and f_2 have the same (opposite) monotonicity.

For example, in the composite function $\ln[x(1-x^2)^{-1/2}]$, the function x^2 increases for $x > 0$, but $(1-x^2)$ is decreasing and positive only for $0 < x < 1$. In this restricted range, however, $(1-x^2)^{-1/2}$ increases, as does $x(1-x^2)^{-1/2}$ and, since the logarithmic function increases, the composite function increases too.

2.6.   PROBLEM SIMPLIFICATION

In the constraint-bound pressure vessel example, monotonicity analysis completely solved the problem. It did this by proving that exactly as many constraints were satisfied as equalities at the optimum as there were design variables to be chosen. Even when this happy situation does not occur, monotonicity analysis is still worth trying. A constraint identified as *active*, i.e., satisfied as a strict equality at the optimum, can often be used to eliminate a design variable and simplify the problem. In the example to follow, a four-variable constrained optimization problem is reduced by monotonicity analysis to that of minimizing an unconstrained function of a single variable. This reduced problem, which is not solvable by monotonicity analysis, is finished off in Chapter 3, which deals with optimizing unconstrained functions.

This new example is based on a cofferdam design problem of Neghabat and Stark, who solved it with the more complicated technique of geometric programming to be described in the next chapter. Simple monotonicity analysis almost manages to complete the task by itself.

A cofferdam is a temporary structure enclosing an ordinarily submerged area to permit construction of a permanent structure there; for example, a bridge pier in the middle of a river. The water level surrounding a cofferdam fluctuates randomly. Any flooding brings costs of repair, replacement, cleaning, and dewatering, as well as possible damage to the permanent structure being built. Raising the height of the cofferdam reduces chances of flooding, at an increased cost of construction. Hence the designed height should be an appropriate balance between construction expense and flood-risk cost.

Cofferdams tend to be designed now by trial and error, the engineer manipulating some design based on experience to take advantage of site conditions while checking for structural stability. Such designs must be feasible, but they are seldom optimal, according to Neghabat and Stark. Their mathematical description, used here, is for circular cellular coffer-

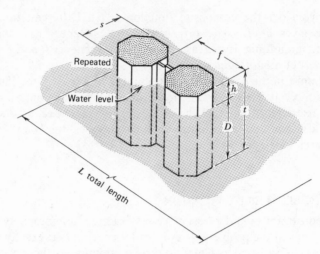

Figure 2-3   Cells of a cofferdam

dams made of earth fill and sheet steel piling resting on rock. The concepts are valid, however, for other styles.

Figure 2-3 shows a schematic arrangement of a cofferdam with its four design variables (total piling length t, cell face length f, cell side length s, and height above low water h) and two geometric parameters (total length L and low-water depth D).

Total cost, to be minimized wrt t, f, s, and h, is the sum of excavation and fill cost, piling face cost, piling side cost, and expected flooding cost. Excavation and fill cost is proportional to the total cofferdam volume, which for fixed total length is proportional to cell side area, and is written Vst with V a parameter. Piling face cost depends directly on the total height t and is written Tt. Piling side cost is proportional to cell side area st as well as to the number of cells, in turn inversely proportional to face length f, giving $Astf^{-1}$. Finally, expected flooding cost has been assumed by Neghabat and Stark to vary inversely with h, the height above low water, giving a cost of Fh^{-1}. Thus the total cost is

$$c = Vst + Tt + Astf^{-1} + Fh^{-1}$$

This increases wrt s and t and decreases wrt f and h, so constraints are essential to bound the cost below.

The first constraint is that when the water is at the top of the cofferdam, exerting maximum force on it, the cell sides must be long enough to keep the pilings from slipping. This gives

$$s \geqslant St$$

Also, the frontal area of each cell must be small enough to prevent yielding to the pressure of the fill.

$$ft \leqslant P$$

Finally, the design variables are related by the geometry in that the total height t is the sum of low-water depth D and the height above low water h.

$$t = h + D$$

The reader can use the monotonicity arguments explained in the pressure vessel problem to prove that both inequalities must be active (i.e., satisfied as strict equalities) at the minimum cost. Moreover, the equation can be replaced by an inequality. All this is summarized using the notation $\leq$ to indicate that all designs satisfying strict inequality $<$ are feasible but not optimal when they are derived from inequalities $\leqslant$. If the original constraint was a strict equality, $\leq$ means that the infeasible designs satisfying $<$ would never be optimal anyway.

$$s \geq St \qquad ft \leq P \qquad t \geq h + D$$

These three equations in four unknowns can be solved for any three variables as functions of the fourth. For reasons to become clear in Sec. 3.3, let this fourth variable be h, so that

$$s \geq S(h+D) \qquad f \leq P(h+D)^{-1} \qquad t \geq h+D$$

Elimination of s, f, and t from the cost gives a function only of h.

$$Vst = VS(h+D)^2 = VSD^2 + 2VSDh + VSh^2$$
$$Tt = T(h+D) = TD + Th$$
$$Astf^{-1} = ASP^{-1}(h+D)^3 = ASP^{-1}D^3 +$$
$$\qquad + 3ASP^{-1}D^2h + 3ASP^{-1}Dh^2 + ASP^{-1}h^3$$
$$c = (VSD^2 + TD + ASP^{-1}D^3) + (2VSD + T + 3ASP^{-1}D^2)h +$$
$$\qquad + (VS + 3ASP^{-1}D)h^2 + ASP^{-1}h^3$$

The constant term is large, being the cost of building the cofferdam up to the low-water level, where $h = 0$. Neghabat and Stark used the parameter values $V = 168$, $T = A = 3660$, $S = 1.0425$, $P^{-1} = 0.00035$, $D = 41.5$, and $F = 50,000$, so

$$VSD^2 + TD + ASP^{-1}D^3 = 549.0(10^3) \equiv \bar{c}$$

Then if the *variable cost* v is defined as

$$v \equiv c - \bar{c}$$

the problem is now to minimize

$$v = (2VSD + T + 3ASP^{-1}D^2)h + (VS + 3ASP^{-1}D)h^2 + ASP^{-1}h^3 + Fh^{-1}$$
$$\equiv C_1 h + C_2 h^2 + C_3 h^3 + Fh^{-1}$$

where C_1, C_2, and C_3 are parameters. This function, being the sum of an increasing function $(C_1 h + C_2 h^2 + C_3 h^3)$ and a decreasing function (Fh^{-1}), is neither increasing nor decreasing for positive h. Clearly $v > 0$ for $h \geqslant 0$, so $v(h)$ is bounded below and hence has a minimum.

Thus monotonicity analysis, by showing that all three constraints must be active, reduced the original four variable problem to one in a single variable. This unconstrained function will be optimized in the next chapter.

2.7. CONCLUDING SUMMARY

Engineering devices tend strongly to be shaped by physical constraints. When these specifications and restrictions are monotonic in the design variables, further monotonicity of the objective function can often show the designer which constraints are active at the optimum, leading to important simplifications. In extreme "constraint bound" cases there is an active constraint for every variable, and the optimum design can be found by solving the constraints without reference to the objective. The monotonicity analysis introduced here is extended in Chapter 5 to situations where it is inconvenient or impossible to solve the active constraints. Although monotonicity analysis can handle simple versions of problems with standard sizes, Chapter 7 gives more powerful methods.

NOTES AND REFERENCES

Partial optimization is a modern name for what must be an ancient mental process. Its expression in terms of functional equations was used in R. Bellman's "dynamic programming" for a particular class of serial optimization problems. The more general term "partial optimization" occurs in *Foundations of Optimization*.

R. C. Buck's calculus book has a development of the Implicit Function Theorem.

The style of reasoning designated here as "monotonicity analysis" is also venerable. It is used verbally in the earliest optimal design books (Johnson, Fox, Middendorf). Duffin, Peterson, and Zener have used it informally in setting up engineering design problems for solution by geometric programming, a topic developed in the next chapters. In his dissertation, Passy conceived rules for directing inequalities that were eventually published as precursors of the present monotonicity theory by Blau.

Blau, G. E., and D. J. Wilde, "Optimal System Design by Generalized Polynomial Programming,' Can. J. Chem. Eng., **47**, 317–326 (1969).

Buck, R. C., Advanced Calculus, 2nd Ed., McGraw-Hill, New York, 1965.

Neghabat, F., and R. M. Stark, "A Cofferdam Design Optimization," Math. Prog., **3**, 263–275 (1972).

Passy, U., Generalization of Geometric Programming; Partial Control of Linear Inventory Systems, Ph.D. dissertation, Stanford, 1966.

Unklesbay, K., G. E. Staats, and D. L. Creighton, "Optimal Design of Pressure Vessels," ASME Paper 72-PVP-2 (1972).

References for other authors are given at the end of Chapter 1.

EXERCISES

2-1. Verify the pressure vessel design of Unklesbay, Staats, and Creighton (Sec. 2.1).

2-2. Derive the procedure for designing a hemispherical head pressure vessel (Sec. 2.1).

2-3. Give two devices, one in and one outside your specialty, that are constraint bound (Sec. 2.2).

2-4. Write the monotonicity analysis for one of the devices of Exercise 2-3 (Sec. 2.3).

2-5. Suppose that pressure vessel plates must be multiples of 6 in. in length and 0.25 in. in thickness, and that the radius to which they can be rolled must be a multiple of 6 in. Use the method of Sec. 2.4 to design a flat head vessel fulfilling the specifications of Unklesbay, Staats, and Creighton.

2-6. Determine the monotonicity, or lack of it, for each variable (Sec. 2.5), together with the range of applicability.
  (a) $\exp[x^{-1}(1-x^2)]^{1/2}$
  (b) $\sin(x^2 + \ln y^{-1})$
  (c) $\exp(-x^2)$

*2-7. Apply monotonicity analysis to your design project to simplify it.

*Project problem.

2-8.   (Washington Braga, Pontifícia Universidade Católica do Rio de Janeiro, Brasil) A cubical refrigerated van is to be designed to transport fruit between São Paulo and Rio. Let n be the number of trips, s the length of a side (cm), a the surface area (cm^2), v the volume (cm^3), and t the insulation thickness (cm). Transportation and labor cost is $21n$; material cost is $16(10^{-4})a$; refrigeration cost is $17(10^{-4})an/(t+1.2)$; and insulation cost is $41(10^{-5})at$—all in April 1977 cruzeiros. The total volume of fruit to be transported is $34(10^6)$ cm^3. Express the problem of designing this van for minimum cost as a constrained optimization problem. Where possible, replace equalities by active inequalities, and determine which inequalities are active at the minimum. Is this system constraint bound? Is it unbounded?

CHAPTER **3**

Estimates
and Bounds

He that knoweth when he hath enough is no fool.

John Heywood (1546)

Engineering practice involves estimating such things as cost, weight, size, or performance. Experienced designers usually express an estimate as a bound rather than an exact figure. The engineer who estimates a million dollars for a piece of equipment may mean that it can be purchased for no more than a million, leaving open the possibility that it may cost less. A performance guarantee involves a lower bound, not an exact prediction, on how well a device operates. Occasionally estimates may have both upper and lower bounds, as when the designer predicts a boiler rating within 10%. A more subtle form of hedging an estimate is to give it in "round numbers," that is, to only one or two significant figures.

Using bounds rather than exact numbers is not evasive; it is merely realistic. Performance of engineering devices is rarely predictable any closer than 10 or 20 %, and demands on equipment may fluctuate wildly. Cost information, being subject to economic variation and commercial manipulation, is even less reliable. So even the designer who uses a calculator carrying 10 significant figures must, to avoid misleading the client, limit the final estimate to one or two. To be fully professional, the engineer should in principle even make clear whether the number given is an upper or a lower bound.

Viewing estimates as bounds and inequalities rather than exact numbers necessarily influences the designer's attitude toward optimization. An elaborate computer study to find the exact optimum to 10 figures would be

55

clearly an exercise in futile perfectionism if none of the data had better than three figure accuracy. Under these circumstances any of a number of designs would have performances indistinguishable from the ideal optimum. This chapter shows that often designs that are close enough to optimal for practical purposes can be generated with much less effort than would be needed to find the theoretical optimum. Equally important, the engineer will learn how to tell when a given design is "satisfactory," that is, indistinguishable from the optimum, rendering further computation unnecessary.

The approximate character of estimates does not mean that they cannot be made rigorous. Having recognized that estimation involves inequalities, the designer can construct mathematically precise bounds with little more effort than would be needed by an informed guess. And the more sophisticated methods described will give estimates that not only are rigorous but also are better than the usual ones.

After briefly describing the "semilogarithmic derivative," a concept needed later, we show how to generate, with physically plausible but strictly rigorous reasoning, a fairly good lower bound on the cofferdam cost. A better bound is then constructed by the simplest of engineers' tricks —neglecting small contributions. In the example, this immediately gives a satisfactory design.

The cofferdam example is then modified so that the simple methods preceding still leave an unacceptable gap between the approximate design and the lower bound on the cost. Then a better lower bound is constructed from the very useful inequality, relatively unknown to engineers, between arithmetic and geometric means. This reduces the gap, but not enough, so the lower bound itself is used to construct a better design falling within the threshold of acceptability by what is called the *invariance method*.

The concept of *satisfactory* design, with the associated idea of a *satisfaction threshold*, is developed rigorously. Although this approach is simple, it will be employed throughout the book to convert relatively difficult optimization problems into the construction of computationally easy bounds and estimates.

Lower bounds are especially easy to construct when the cost to be minimized is a sum of power functions, as in the cofferdam problem and many others in engineering design. Such functions are called *posynomials*, and the theory of their optimization is called *geometric programming*. It is especially interesting when the number of terms exceeds the number of variables by exactly one; in this case the optimal solution is given by closed form formulas of the kind populating design handbooks. This will be demonstrated in the derivation of a procedure for allocating tolerances among machine parts.

Interesting design rules can be discovered even when there are more terms than this, as illustrated in a pipeline design problem. Three ways of generating improved designs are presented: condensation, partial invariance, and dual geometric programming.

The construction of bounds rather than the pursuit of an ideal optimum will be a theme running throughout the book. It is not just a way to save effort, although it usually does. More important, it is a conservative safeguard against falsely accepting an inferior design which might superficially appear optimal.

3.1. SEMILOGARITHMIC DERIVATIVES

Consider again the variable cost v of the cofferdam problem posed in the preceding chapter. If the expression for $\partial v / \partial h$ is multiplied by h, the resulting expression is a linear function of the original terms of the objective. To see this, define $t_1 \equiv C_1 h$; $t_2 \equiv C_2 h^2$; $t_3 \equiv C_3 h^3$; $t_4 \equiv Fh^{-1}$; so that $v = t_1 + t_2 + t_3 + t_4$. Then

$$h\left(\frac{\partial t_1}{\partial h}\right) = h(C_1) = t_1$$

$$h\left(\frac{\partial t_2}{\partial h}\right) = h(2C_2 h) = 2t_2$$

$$h\left(\frac{\partial t_3}{\partial h}\right) = h(3C_3 h^2) = 3t_3$$

$$h\left(\frac{\partial t_4}{\partial h}\right) = h(-Fh^{-2}) = -t_4$$

In general, if $t \equiv Cx^\alpha$, where α and C are any real numbers, then

$$x\left(\frac{\partial t}{\partial x}\right) = x(\alpha Cx^{\alpha-1}) = \alpha t.$$

The function

$$x\left(\frac{\partial t}{\partial x}\right) \equiv \frac{\partial t}{\partial \ln x}$$

is called the *first semilogarithmic derivative* of t wrt x, or "semilog derivative" for short. The semilog derivative is useful because its differentiation formula is so simple for the power function Cx^α arising so often in engineering and economics. Notice that for positive x, the first semilog derivative has the same sign as the first derivative, and of course both types of derivative vanish at the same values of x. Given the numerical values of the terms t, which must be computed anyway when estimates are being made, it is easy to calculate their semilog derivatives.

The *second* semilog derivative is also easy to calculate from the numerical value of a power function term. Since the first semilog derivative of a power function is still a power function, another semilog differentiation yields a power function again. More important, the second semilog derivative, like the first, is proportional to the original function. Thus if $t = Cx^\alpha$,

$$\frac{\partial^2 t}{\partial (\ln x)^2} = \frac{\partial}{\partial \ln x}\left(\frac{\partial t}{\partial \ln x}\right) = \frac{\partial}{\partial \ln x}(\alpha t) = \alpha \frac{\partial t}{\partial \ln x} = \alpha^2 t$$

In the cofferdam example,

$$\frac{\partial v}{\partial \ln h} = t_1 + 2t_2 + 3t_3 - t_4$$

$$\frac{\partial^2 v}{\partial (\ln h)^2} = t_1 + 4t_2 + 9t_3 + t_4$$

This shows the second semilog derivative is positive for all positive h. Since v itself must be positive in this range, it is bounded below by zero and therefore has a minimum. The positivity of $\partial^2 v / \partial (\ln h)^2$ ensures that there is only one value of h where $\partial v / \partial \ln h$ vanishes, and that must be at the global minimum.

3.2. POSYNOMIALS

The cofferdam cost function is an example of an important class of functions occurring quite frequently in engineering-design problems. This cost is a sum of positive terms, each of which is a power function. In general, such functions are called *posynomials*, an adaptation of the word "polynomial" to the case where all coefficients are positive. Note, however, that the power-function exponents, which must be positive integers for *poly*nomials, can be any real number for *posy*nomials.

The properties observed in the cofferdam example are generally true for all posynomials of one nonnegative variable. That is, all of them have positive second semilog derivatives, indicating that there is no more than one point where the first semilog derivative can vanish. If the posynomial neither strictly increases (in which case it would have all exponents positive) nor strictly decreases (in which case it would have all exponents negative), then the global minimum occurs at the unique point where the first semilog derivative vanishes. The independent variable will be positive at the minimum. In Sec. 3.10 these properties are proven to extend to posynomials in more than one variable.

3.3. SIMPLE LOWER BOUNDS

It would be easy, in the simple cofferdam example, to find the global minimum with great accuracy and little effort by scanning the function directly. But this approach soon becomes ineffective as the number of variables increases. Consequently, the foundation will now be lain for methods promising to be useful in more complicated situations. The approach to be developed is not to seek the minimum, but rather to establish a lower bound on it.

This procedure motivated the choice of h rather than t as the design variable remaining in the cofferdam cost function. In this way, the total cost c could be bounded below, for

$$c(h) \equiv \bar{c} + v(h) > \bar{c} = 549(10^3)$$

This lower bound is important to the designer, for it says that no cofferdam can do the job for less than \$549,000. Thus a contractor bidding for the construction had better stay higher or else not go after the contract. The economic interpretation of the lower bound $\bar{c}$, which equals the construction cost $c(0)$ when $h=0$, is the cost of a cofferdam built only up to the low-water level. Obviously, a realistic design must cost more. This plain fact would have been obscured, however, if the total piling length t had been the independent variable instead of h, for then the constant $\bar{c}$ would not have emerged from the elimination of the variables remaining.

Next consider ways of bounding the variable cost v, which has already been shown to be strictly positive for $h \geqslant 0$. The first things to look for are terms small enough to neglect in a first approximation. To do this, one must have numerical values of the coefficients, and for the cofferdam,

$$v = 25,100h + 341h^2 + 1.34h^3 + 50,000h^{-1}$$

The first term $25,100h$ will be greater than the second $341h^2$ for $h <$ $25,100(341)^{-1} = 73$, and it will exceed the third $1.34h^2$ for $h <$

$(25,100(1.34)^{-1})^{1/2} = 136$. Since in this problem the length D of the pilings below low water is 41.5, it would be reasonable to assume h will be much smaller than 73 at the optimum, and so a *lower-bounding function* is constructed by neglecting the second and third terms.

$$v(h) > 25,100h + 50,000h^{-1} \equiv l(h)$$

The right member is called a lower-bounding *function* rather than a lower *bound*, because it still depends on the independent variable h. Nevertheless the inequality holds throughout the range of h.

To get a lower *bound*, that is, a *constant* that bounds $v(h)$ below, it would be reasonable to seek the minimum value l_* of the lower-bounding function. The first semilog derivative of $l(h)$ is $\dfrac{\partial l}{\partial \ln h} = t_1 - t_4$. This is just the difference between the first and fourth terms of $v(h)$. Thus $l(h)$ is at its minimum when this difference is zero, i.e., when the terms are equal:

$$t_1 = t_4$$

This equality can be interpreted as an interesting design rule. But before discussing this aspect, let us confirm that it gives a precise, unique, unambiguous design. As functions of h, the two cost terms are $t_1 \equiv C_1 h = t_4 \equiv Fh^{-1}$. If $\bar{h}$ represents the solution to this equation, then l is minimum when

$$\bar{h} = (F/C_1)^{1/2} \equiv [F/(2VSD + T + 3ASP^{-1}D^2)]^{1/2}$$
$$= (50,000/25,100)^{1/2} = 1.41$$

The minimum value l_* of $l(h)$ is

$$l_* = l(\bar{h}) = t_1 + t_4 = 25,100\bar{h} + 50,000\bar{h}^{-1} = 70.9(10^3)$$

Hence the total cost is bounded below by

$$c(h) = \bar{c} + v(\bar{h}) > 549.0(10^3) + 70.9(10^3) = 619.9(10^3)$$

This tells the designer that no design can cost less than $619.9(10^3)$. The true total cost of the design corresponding to $\bar{h}$ is obtained by adding in the cost of the two neglected terms.

$$c(\bar{h}) = 619.9(10^3) + C_2\bar{h}^2 + C_3\bar{h}^3 = 619.9(10^3) + 679 + 4$$
$$= (619.9 + 0.7)(10^3) = 620.6(10^3)$$

This is an *upper* bound on the as yet unknown minimum value $c_\star$ of $c(h)$, since by definition, the minimum must be less. So $c_\star$ is now known to be in the narrow range $620.6(10^3) \geqslant c_\star \geqslant 619.9(10^3)$. Now the original model is certainly accurate to no more than two significant figures, if that. Hence, more realistically,

$$62(10^4) \geqslant c_\star \geqslant 62(10^4)$$

This means that $\bar{h}$, although maybe not the true minimizing value, gives a value of the cost function that cannot be distinguished from the true minimum, given the imprecision of the model. Therefore the search for the minimum cost design can be terminated at $\bar{h}$, at least for the values of the design parameters given. The value of the lower bound $l_\star$ depends entirely on the design parameters, since

$$l_\star = C_1\bar{h} + F\bar{h}^{-1} = C_1(F/C_1)^{1/2} + F(F/C_1)^{-1/2} = 2(C_1F)^{1/2}$$
$$= 2\left[(2VSD + T + 3ASP^{-1}D^2)F\right]^{1/2}$$

Remarkably, an estimator can say now with certainty that the total construction cost, excluding flooding cost, will be at least $\bar{c} + 2(C_1F)^{1/2}$, which has been shown to be $\$62(10^4)$.

Since the rough design $\bar{h}$ got very close to the optimum, it is well to develop a simple rule for finding it. The equality of the terms t_1 and t_4 can be expressed as a design rule:

Design the cofferdam so that the (linear) construction cost equals the expected flooding cost.

The linear construction cost is defined as the cost with all terms neglected of order higher than first.

In practice the total length t $(= D + h)$ must be a standard size. The designer then needs to know whether to round up or down on the quantity $D + \bar{h}$. The semilog derivative at $\bar{h}$ is

$$t_1 + 2t_2 + 3t_3 - t_4 = 2t_2 + 3t_3 > 0$$

since $t_1 - t_4 = 0$ at $\bar{h}$. Hence $v(h)$ increases with h in the neighborhood of $\bar{h}$, and so $h_\star < \bar{h}$ and $t < D + \bar{h}$. It would be reasonable then to check the design for t rounded down, i.e., $\lfloor t \rfloor \equiv \lfloor D + \bar{h} \rfloor$. If the variable cost for this design is close enough to the lower bound $l_\star$, then no further study is needed. Otherwise it may be wise to check the design for t rounded *up* to the next larger standard size $\lceil t \rceil \equiv \lceil D + \bar{h} \rceil$. The design with the lesser cost will, of course, be the optimal *standard* design, whether or not it deviates noticeably from the lower bound $l_\star$.

In the Neghabat and Stark example, the first approximation came so close that no further improvement was needed. As an extreme test of the approximation, as well as a demonstration of a method for improving a design, consider the cofferdam problem with $D=0$ instead of 41.5. Then $\bar{c}=0$, $c=v$, and

$$v = 3660h + 175h^2 + 1.34h^3 + 50,000h^{-1}$$

$$> 2\left[(3660)(50,000) \right]^{1/2} = 27.0(10^3)$$

$$\bar{h} = \left(\frac{50,000}{3660} \right)^{1/2} = 3.70$$

$$v(\bar{h}) = 10^3\left[27 + 175(3.70)^2 + 1.34(3.70)^3 \right] = 29.5(10^3)$$

Thus

$$29.5(10^3) \geqslant v_\star > 27.0(10^3)$$

This gap of $\$2.5(10^3)$ is large enough to justify further effort. At this point it is unclear whether the gap exists because v can be reduced further, or because the lower bound is not sharp enough. In the three sections that follow, we develop a way to find a better lower bound. Then the new theory is used to produce an improved design by what will be called the *invariance* method.

3.4. GEOMETRIC MEAN

In Sec. 3.5 we show how to construct a lower bound greater, and hence better, than the one already available. This requires a result to be developed here involving various ways of averaging positive real numbers. The objective function will be one of these averages and the lower bound another.

Consider Fig. 3-1 showing a rectangle of length l and height h, together with an s by s square having the same area A as the rectangle. Then $A = lh = s^2$ and $s = l^{1/2}h^{1/2}$. The quantity $l^{1/2}h^{1/2}$ is called the *geometric mean* of l and h, a definition holding for any pair of positive numbers. In words, the geometric mean of the length and height of a rectangle equals the side of a square having the same area.

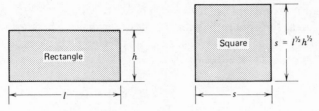

Figure 3-1 Geometric mean

The more familiar *arithmetic mean*, commonly called the *average*, of the length and height is $(l/2)+(h/2)$. Thus the perimeter p of the rectangle is four times the *arithmetic* mean.

$$p=4\big[(l/2)+(h/2)\big]$$

The perimeter of the square having the same area is, of course, four times the *geometric* mean.

$$4s=4l^{1/2}h^{1/2}$$

The Ancient Greeks posed the question: Among all rectangles having the same area, which has the least perimeter? The answer can be obtained by the following purely algebraic reasoning, requiring no calculus, even though the problem does involve optimization.

$$p=4\left(\frac{l}{2}+\frac{h}{2}\right)=4\left[\left(\frac{l}{2}+\frac{h}{2}\right)^2\right]^{1/2}$$

$$=4\left(\frac{lh}{2}+\frac{l^2}{4}+\frac{h^2}{4}\right)^{1/2}$$

$$=4\left(lh-\frac{lh}{2}+\frac{l^2}{4}+\frac{h^2}{4}\right)^{1/2}$$

$$=4\left[lh+\left(\frac{l}{2}-\frac{h}{2}\right)^2\right]^{1/2}\geq 4l^{1/2}h^{1/2}\equiv 4s$$

because

$$\left(\frac{l}{2}-\frac{h}{2}\right)^2\geq 0$$

with equality if and only if $l = h = s$. Therefore among all rectangles with the same area, the square has the smallest perimeter. In other words, the arithmetic mean side length $(l + h)/2$ of any rectangle is never less than the geometric mean side length $(lh)^{1/2}$.

$$\frac{l+h}{2} \geq (lh)^{1/2}$$

The arithmetic and geometric means are equal only for the square, in which $l = h = s$.

Suppose now that the vertical sides of the rectangle cost three times as much as the horizontal sides, giving a total cost proportional to $c \equiv 3h + l$. A little algebra gives the proportions minimizing this average cost for a given area. Since the cost function increases in both h and l, the area constraint $hl = A$ can be written as an inequality known to be tight at the minimum.

$$hl \geq A$$

This bounds both h and l below.

The cost can be written as an arithmetic mean.

$$c = 3h + l = \tfrac{1}{2}(6h) + \tfrac{1}{2}(2l)$$

This is bounded below by the geometric mean of the positive quantities $6h$ and $2l$.

$$c \geq (6h)^{1/2}(2l)^{1/2} = (12)^{1/2}h^{1/2}l^{1/2}$$

The constraint gives a lower bound on the right member.

$$(12)^{1/2}h^{1/2}l^{1/2} \geq (12)^{1/2}A^{1/2}$$

Therefore

$$c \geq (12)^{1/2}A^{1/2}$$

Since strict equality can be achieved, the minimum cost is $(12)^{1/2}A^{1/2}$, computed without knowing the optimal design.

$$c_\star = (12)^{1/2}A^{1/2}$$

The optimum height $h_\star$ can be found by realizing that the inequalities become active only when the positive quantities $6h$ and $2l$ are equal.

Hence $6h_\star = 2l_\star$, or $l_\star / h_\star = 3$. This gives the proportions of the optimal design. To obtain the optimizing values, use the fact that each cost term must be half the total. Then $6h_\star = \frac{1}{2}c_\star = \frac{1}{2}(12)^{1/2}A^{1/2} = 3^{1/2}A^{1/2}$ so that $h_\star = A^{1/2}\sqrt{3}/6$ and $l_\star = 3h_\star = A^{1/2}\sqrt{3}/2$. A more convenient design rule is that the minimum cost design has both horizontal and vertical costs equal, which holds for all cost ratios.

Consider now two positive variables x_1 and x_2. The function $\frac{3}{4}x_1 + \frac{1}{4}x_2$ is called a *weighted arithmetic mean* of x_1 and x_2, with weights $\frac{3}{4}$ and $\frac{1}{4}$. Weights must always sum to unity. The following algebra leads to the concept of weighted *geometric* mean.

$$\tfrac{3}{4}x_1 + \tfrac{1}{4}x_2 = \tfrac{1}{2}x_1 + \left(\tfrac{1}{4}x_1 + \tfrac{1}{4}x_2\right) = \tfrac{1}{2}x_1 + \tfrac{1}{2}\left(\tfrac{1}{2}x_1 + \tfrac{1}{2}x_2\right)$$

$$\geq \tfrac{1}{2}x_1 + \tfrac{1}{2}x_1^{1/2}x_2^{1/2} \geq x_1^{1/2}\left(x_1^{1/2}x_2^{1/2}\right)^{1/2} = x_1^{3/4}x_2^{1/4}$$

The right member $x_1^{3/4}x_2^{1/4}$ is called the *weighted geometric mean* of x_1 and x_2 with weights $\frac{3}{4}$ and $\frac{1}{4}$. Thus in this case the weighted arithmetic mean is bounded below by the weighted geometric mean, the weights being the same in both cases.

Next consider an arithmetic mean of three variables, with equal weights.

$$\tfrac{1}{3}x_1 + \tfrac{1}{3}x_2 + \tfrac{1}{3}x_3 = \tfrac{1}{3}x_1 + \tfrac{2}{3}\left(\tfrac{1}{2}x_2 + \tfrac{1}{2}x_3\right)$$

$$\geq \tfrac{1}{3}x_1 + \tfrac{2}{3}x_2^{1/2}x_3^{1/2} \geq x_1^{1/3}\left(x_2^{1/2}x_3^{1/2}\right)^{2/3} = x_1^{1/3}x_2^{1/3}x_3^{1/3}$$

Again the arithmetic mean of three variables is never less than the corresponding geometric mean. It can be proven by induction that

$$x_1/n + x_2/n + \cdots + x_n/n \geq x_1^{1/n}x_2^{1/n}\cdots x_n^{1/n}$$

or

$$\sum_{i=1}^{n} x_i/n \geq \prod_{i=1}^{n} x_i^{1/n}$$

Equality is achieved if and only if $x_1 = x_2 = \cdots = x_n$.

More generally, the weighted geometric mean always bounds the weighted arithmetic mean below. Formally, let $w_1, w_2, \ldots, w_n$ be n positive weights summing to unity. Then weighted arithmetic mean $\equiv \sum_{i=1}^{n} w_i x_i \geq \prod_{i=1}^{n} x_i^{w_i} \equiv$ weighted geometric mean with equality if and only if all x_i are equal. This important relation, called the *geometric inequality*, will now be used to obtain a better lower bound on the cofferdam cost function.

3.5.   GEOMETRIC LOWER-BOUNDING FUNCTION

Return now to the cofferdam example. The first approximate design was $\bar{h} = 3.70$, for which the variable cost was

$$v = (3.66h + 0.175h^2 + 0.00134h^3 + 50.0h^{-1})10^3$$

The terms of $v(\bar{h})$ are

$$v(\bar{h}) = (13.53 + 2.39 + 0.07 + 13.53)10^3 = 29.51(10^3) \equiv t_1 + t_2 + t_3 + t_4$$

Weights for these four costs can be defined by dividing the terms by the total cost.

$$w_i \equiv t_i / v$$

Then $w_1 = 0.458$; $w_2 = 0.081$; $w_3 = 0.002$; $w_4 = 0.458$. Write $v(h)$ as a weighted arithmetic mean, using these weights.

$$v = \left[0.458\left(\frac{3.66h}{0.458} \right) + 0.081\left(\frac{0.175h^2}{0.081} \right) \right.$$
$$\left. + 0.002\left(\frac{0.00134h^3}{0.002} \right) + 0.458\left(\frac{50.0h^{-1}}{0.458} \right) \right]10^3$$

This is bounded below by the weighted geometric mean.

$$(10^{-3})v \geq \left(\frac{3.66h}{0.458} \right)^{0.458}\left(\frac{0.175h^2}{0.081} \right)^{0.081}\left(\frac{0.00134h^3}{0.002} \right)^{0.002}\left(\frac{50.0h^{-1}}{0.458} \right)^{0.458}$$
$$= 23.7h^{0.169}$$

The right member is a valid lower-bounding function for v. But this is no help, because its minimum value is zero, being an increasing function of h. What is needed is a constant lower bound, not a lower-bounding function.

3.6.   GEOMETRIC LOWER-BOUNDING CONSTANT

The geometric inequality yielded a lower-bounding *function*, rather than the desired constant, because the weights were improperly chosen. After all, weights can be any numbers summing to unity, not just those generated by the costs for a given design. To emphasize this distinction, use new symbols ω_1, ω_2, ω_3, and ω_4 (read "omega one," etc.) to represent weights for constructing arithmetic and geometric means. They must sum to unity.

$$\omega_1 + \omega_2 + \omega_3 + \omega_4 = 1$$

Moreover, they must be chosen to make the geometric mean a constant rather than a function of the design variable h. This requires another restriction, which we derive now.

Write the objective as a weighted arithmetic mean, using the new weights ω_1, ω_2, ω_3, and ω_4.

$$v(10^{-3}) = \omega_1 \left(\frac{3.66h}{\omega_1} \right) + \omega_2 \left(\frac{0.175h^2}{\omega_2} \right) + \omega_3 \left(\frac{0.00134h^3}{\omega_3} \right) + \omega_4 \left(\frac{50.0h^{-1}}{\omega_4} \right)$$

This is bounded below by the geometric mean of the same numbers.

$$v(10^{-3}) \geq \left(\frac{3.66h}{\omega_1} \right)^{\omega_1} \left(\frac{0.175h^2}{\omega_2} \right)^{\omega_2} \left(\frac{0.00134h^3}{\omega_3} \right)^{\omega_3} \left(\frac{50.0h^{-1}}{\omega_4} \right)^{\omega_4}$$

$$= \left(\frac{3.66}{\omega_1} \right)^{\omega_1} \left(\frac{0.175}{\omega_2} \right)^{\omega_2} \left(\frac{0.00134}{\omega_3} \right)^{\omega_3} \left(\frac{50.0}{\omega_4} \right)^{\omega_4} h^{\omega_1 + 2\omega_2 + 3\omega_3 - \omega_4}$$

To make this expression a constant independent of h, the exponent of h must be set to zero, giving the restriction

$$\omega_1 + 2\omega_2 + 3\omega_3 - \omega_4 = 0$$

There are now two linear equations relating the four weights. Hence if any pair of weights are given numerical values, the other two are uniquely determined. For example, if costs t_2 and t_3 are neglected as in the earlier approximation, which corresponds to setting $\omega_2 = \omega_3 = 0$, then $\omega_1 + \omega_4 = 1$; $\omega_1 - \omega_4 = 0$, for which the solution is $\omega_1 = \omega_4 = \frac{1}{2}$. This of course is precisely the result obtained earlier, namely, that the linear construction cost equals the flooding cost in this approximation.

Setting weights to zero appears at first glance to cause difficulties in terms such as $\omega_2(0.175h^2/\omega_2)$ in the arithmetic mean, and factors such as $(0.175/\omega_2)^{\omega_2}$ in the geometric mean. Remember, however, that setting ω_2 and ω_3 to zero corresponds to deleting the second and third terms from the objective function entirely, leaving only the first and fourth, the sum of which was called $l(h)$ in the earlier context.

$$l(h) \equiv t_1 + t_4 = \omega_1 \left(\frac{3.66h}{\omega_1} \right) + \omega_4 \left(\frac{50.0h^{-1}}{\omega_4} \right)$$

The geometric mean for $l(h)$ does not contain the factors $(0.175h^2/\omega_2)^{\omega_2}$ and $(0.00134h^3/\omega_3)^{\omega_3}$.

$$(10^{-3})l(h) \geq \left(\frac{3.66h}{\omega_1}\right)^{\omega_1}\left(\frac{50.0h^{-1}}{\omega_4}\right)^{\omega_4}$$

Thus a factor in the geometric mean is deleted, or equivalently, set to unity whenever its weight is zero. This is not merely a formal artifice, for it can be shown using L'Hôpital's Rule that if c is a positive constant, then $\lim_{\omega \to 0} (c/\omega)^\omega = 1$. To verify this numerically, make the following computation on your calculator: $(98/10^{-9})^{10^{-9}} = 1.000000025$.

Using $\omega_1 = \omega_4 = \frac{1}{2}$, which cancels h from the geometric mean,

$$(10^{-3})l(h) \geq (3.66/0.5)^{0.5}(50.0/0.5)^{0.5} = 27.06 \equiv l_\star$$

This verifies the lower bound found earlier by other methods. However, this is the set of weights that gave an excessive gap between the upper and lower bounds on $v_\star$.

It is reasonable to expect an improved lower bound by not neglecting t_2 and t_3, which together account for 8.3% of the cost at $\bar{h}$. Thus use $\omega_2 = 0.081$ and $\omega_3 = 0.002$, the true weights at $\bar{h}$, as estimates of ω_2 and ω_3, and then compute ω_1 and ω_4 from

$$\omega_1 + \omega_4 = 1 - \omega_2 - \omega_3 = 1 - w_2 - w_3 = 0.917$$
$$\omega_1 - \omega_4 = -2\omega_2 - 3\omega_3 = -0.169$$

The solution is $\omega_1 = 0.374$ and $\omega_4 = 0.543$. Application of the geometric inequality using these weights gives

$$(10^{-3})v = 0.374\left(\frac{3.66h}{0.374}\right) + 0.081\left(\frac{0.175h^2}{0.081}\right)$$
$$+ 0.002\left(\frac{0.00134h^3}{0.002}\right) + 0.543\left(\frac{50.0h^{-1}}{0.543}\right)$$
$$\geq \left(\frac{3.66}{0.374}\right)^{0.374}\left(\frac{0.174}{0.081}\right)^{0.081}\left(\frac{0.00134}{0.002}\right)^{0.002}\left(\frac{50.0}{0.543}\right)^{0.543}h^0$$
$$= 29.1$$

Since $v(\bar{h}) = 29.5(10^3)$, the interval containing $v_\star$ is only

$$29.5(10^3) \geq v_\star \geq 29.1(10^3)$$

The parameters being known only to two significant figures, further search is not justified. The original gap was due more to imprecision of the lower bound than to inadequacy of the design.

3.7. SATISFACTORY DESIGNS

In the preceding example it was easy to find a design that was satisfactory in the sense of being indistinguishable from the true optimum, given the precision level of the data. This approach, of great utility to designers, is called *satisfaction*. Rigorously speaking, it involves choosing in advance a positive *satisfaction threshold* $\tau > 0$ and then finding any satisfactory design $\bar{x}$, together with a lower bound l such that for all feasible x, $y(x) \geq l$ and $y(\bar{x}) - l \leq \tau$. This can be expressed equivalently as

$$y(x) \geq l \geq y(\bar{x}) - \tau$$

The above definition holds when the optimum sought is a minimum. When, as in Chapter 5, the objective is a profit to be maximized, then the satisfaction problem is to find a satisfactory design $\bar{x}$ and an *upper* bound u such that for all feasible designs x, $y(x) \leq u$ and $u - y(\bar{x}) \leq \tau$, or equivalently,

$$y(x) \leq u \leq y(\bar{x}) + \tau$$

Note that the inequalities are reversed and the sign of the threshold changed compared to the minimization case.

The brevity of this section perhaps understates the importance of the simple idea of satisfaction. But this concept can save a great deal of effort by ending searches for optima that have passed the point of diminishing returns. Euripides could have been speaking to the modern designer when he wrote, two millenia ago:

Enough is abundance to the wise.

3.8. LOGARITHMIC DERIVATIVE

Perceptive readers may already have noticed similarity between the condition needed to make the geometric mean constant

$$\omega_1 + 2\omega_2 + 3\omega_3 - \omega_4 = 0$$

and the semilog derivative expressed as a function of the objective terms.

$$\frac{\partial v}{\partial \ln h} = t_1 + 2t_2 + 3t_3 - t_4$$

In both expressions, the coefficients are the exponents of h in the corresponding terms. An even closer relation is obtained by dividing the semilog derivative by the objective to obtain the *logarithmic derivative* (log derivative for short):

$$v^{-1}\frac{\partial v}{\partial \ln h} \equiv \frac{\partial \ln v}{\partial \ln h}$$

$$= \frac{t_1}{v} + 2\left(\frac{t_2}{v}\right) + 3\left(\frac{t_3}{v}\right) - \frac{t_4}{v}$$

$$= w_1 + 2w_2 + 3w_3 - w_4$$

Thus the log derivative has exactly the same form as the left member of the condition for making the geometric mean constant. In fact, at the minimum, both semilog and log derivatives must be zero, in which case all these conditions are the same.

$$\left(\frac{\partial v}{\partial \ln h}\right)_* = \left(\frac{\partial \ln v}{\partial \ln h}\right)_* = 0$$

To preserve the subtle distinctions between these similar relations, let the condition on the terms

$$\left(\frac{\partial v}{\partial \ln h}\right)_* = 0$$

be called the *term optimality condition*; the condition on the weights,

$$\left(\frac{\partial \ln v}{\partial \ln h}\right)_* = 0$$

the *weight optimality condition*; and the condition for making the geometric mean constant, for example,

$$\omega_1 + 2\omega_2 + 3\omega_3 - \omega_4 = 0$$

the *orthogonality condition*. Notice that in the example, as will be true in general, the values of the terms and weights satisfying the optimality conditions are unique. However, there is an infinity of weights ω_1, ω_2, ω_3, and ω_4 that can satisfy the orthogonality condition. To distinguish between the **w** and the ω, the former will be called *primal* weights, whereas the latter will be called *dual* weights. Thus the primal weights correspond to real, although not necessarily optimal, designs: $w_i \equiv t_i/v$. On the other hand, the dual weights are any nonnegative solutions to the orthogonality conditions that sum to unity. Usually, dual weights do *not* correspond to real designs. At the minimum, and nowhere else, the *optimal* primal weights $\mathbf{w}_\star$ and the *optimal* dual weights $\omega_\star$ are equal: $\mathbf{w}_\star = \omega_\star$.

3.9. INVARIANCE

The arithmetic mean equals the geometric mean if and only if all the variables are equal. In the example, this means

$$\frac{3.66h}{\omega_{1\star}} = \frac{0.175h^2}{\omega_{2\star}} = \frac{0.00134h^3}{\omega_{3\star}} = \frac{50.0h^{-1}}{\omega_{4\star}} = v_\star(10^{-3})$$

These relations, together called the *Invariance Principle*, can generate an improved design from a lower-bound computation, as will now be demonstrated.

Recall that at the first design $\bar{h} = 3.70$, the variable cost $v(\bar{h}) = 29.5(10^3)$. The lower bound, computed for

$$\omega_1 = 0.374, \ \omega_2 = 0.081, \ \omega_3 = 0.002, \ \omega_4 = 0.543, \text{ is } 29.1(10^3).$$

Let these dual weights be used to find h satisfying the invariance conditions. For the first term,

$$3.66h/0.374 = 39.1 \text{ gives } h = 2.79$$

The other 3 estimates of h are

$$0.175h^2/0.081 = 29.1 \text{ gives } h = 3.67$$
$$0.0013h^3/0.002 = 29.1 \text{ gives } h = 3.68$$
$$50.0h^{-1}/0.543 = 29.1 \text{ gives } h = 3.17$$

These four estimates of h are different because the dual weights are not optimal. Take the average

$$\hat{h} = (2.97 + 3.67 + 3.68 + 3.17)/4 = 3.37$$

The variable cost for this new design is $v(\hat{h}) = 29.2(10^3)$ which is better, at least in the third significant figure, than the approximate design at $\bar{h} = 3.70$. Thus the minimum cost can be bracketed closely by dual weights, not only from below, but also from above.

This *invariance method*, although it worked well in this example, must be used cautiously, since it can produce a design worse than the original one. Good insurance against this is always to compute the objective function for the new design to see what improvement, if any, has been achieved.

3.10. UNCONSTRAINED GEOMETRIC PROGRAMMING

The preceding example will now be generalized to derive the solution to an important class of engineering problems: those involving an unconstrained posynomial objective. Let there be N independent positive variables $x_1, \ldots, x_N$, with the column vector $(x_1, \ldots, x_N)^T$ abbreviated as $\mathbf{x}$. The objective function posynomial, to be minimized wrt $\mathbf{x}$, is written $p(\mathbf{x})$. It is a sum of T positive power functions of $\mathbf{x}$, each called a *term* $t_t(\mathbf{x})$, where the subscript t ranges from 1 to T to identify the terms.

$$p(\mathbf{x}) \equiv \sum_{t=1}^{T} t_t(\mathbf{x})$$

Being a power function of $\mathbf{x}$, a typical term $t_t(\mathbf{x})$ has the form

$$t_t(\mathbf{x}) \equiv C_t \prod_{n=1}^{N} x_n^{\alpha_{tn}}$$

where C_t is a positive parameter, and the exponents α_{tn} are any real numbers, positive or negative. The *unconstrained geometric programming problem* is to minimize $p(\mathbf{x})$ wrt $\mathbf{x}$. This is also called the *unconstrained posynomial problem*.

The solution will be obtained by constructing a weighted geometric mean of the terms, just as in the example. Thus define a positive *dual weight* ω_t corresponding to each term t_t. These must sum to unity, being weights.

$$\sum_{t=1}^{T} \omega_t = 1$$

This sum is called a *normality condition*.

The geometric inequality gives

$$p(\mathbf{x}) = \sum_{t=1}^{T} \omega_t \left(\omega_t^{-1} C_t \prod_{n=1}^{N} x_n^{\alpha_{tn}} \right) \geq \prod_{t=1}^{T} \left(\omega_t^{-1} C_t \prod_{n=1}^{N} x_n^{\alpha_{tn}} \right)^{\omega_t}$$

$$\equiv d(\omega) \prod_{n=1}^{N} x_n^{\gamma_n(\omega)}$$

where for abbreviation are defined the *dual function*

$$d(\omega) \equiv \prod_{n=1}^{N} (C_t/\omega_t)^{\omega_t}$$

and the *dual exponents*

$$\gamma_n(\omega) \equiv \sum_{t=1}^{T} \alpha_{tn}\omega_t, \qquad n = 1, \ldots, N$$

There is strict equality between the left and right members if and only if

$$\omega_t^{-1} C_t \prod_{n=1}^{N} x_n^{\alpha_{tn}} = p(\mathbf{x}), \qquad t = 1, \ldots, T$$

Rearrangement of these *invariance conditions* shows why the ω_t are called "dual *weights*."

$$\omega_t = \frac{t_t}{p}$$

That is, strict equality holds if and only if every dual weight ω_t equals the fraction of p accounted for by the tth term t_t.

The lower bounding function can be made independent of $\mathbf{x}$ by setting the dual exponents $\gamma_n(\omega)$ to zero.

$$\sum_{t=1}^{T} \alpha_{tn}\omega_t = 0, \qquad n = 1, \ldots, N$$

These linear equations, one for each design variable x_n, are called the *orthogonality conditions*.

There are T dual weights ω_t that must satisfy the N orthogonality conditions as well as the single normality condition. Thus if $T - (N+1)$

dual weights are chosen numerically, say by matching them to the known cost distribution for a given design, then all the remaining dual weights can be computed. In fact, if $T - N - 1 = 0$, as in the first approximation to the cofferdam problem (Sec. 3.3), then all dual weights are completely determined without any arbitrary choices. For this reason, the quantity $T - N - 1$, which is the number of degrees of freedom in the combined orthogonality and normality conditions, is called the number of *degrees of difficulty* (the more there are, the harder the geometric programming problem). As the next section demonstrates, problems with zero degrees of difficulty have elegant closed-form solutions. Moreover, approximations involving zero degrees of difficulty are often used to bound design problems having several degrees of difficulty, as Sec. 3.16 shows.

Whether or not there are degrees of difficulty, let any set of dual weights that satisfy the orthogonality and normality conditions be called *dual variables* and designated δ_t for term t. Then the geometric inequality becomes

$$p(\mathbf{x}) \geq D(\boldsymbol{\delta}) = \prod_{n=1}^{N} \left(\frac{C_t}{\delta_t} \right)^{\delta_t}$$

Since the lower-bounding function no longer depends on $\mathbf{x}$, any choice of dual variables supplies a lower bound on the objective. The various implications of this are developed in the sections immediately following. First, the elegant zero-degree-of-difficulty case is illustrated with two examples. Next, some ways of obtaining satisfactory designs when there is a degree of difficulty are examined. The chapter ends by completing the development started here so that the designer can understand the numerical solution techniques of dual geometric programming, even though these will not be employed in this book.

3.11. ZERO DEGREES OF DIFFICULTY: THE GRAVEL BOX

The geometric programming approach is at its most impressive when there are several independent variables but no degrees of difficulty. This has never been illustrated better than by a problem posed by Duffin who named geometric programming. He considered the design of an open topped box for transporting a pile of material, say gravel, from one place to another. If l, h, and w represent the length, height, and width respectively of the box, then the number of loads is V/lhw, where V is the volume of the pile, the only physical parameter of the problem. Let the cost parameters be the cost of transporting one load T, the cost per unit area of bottom B, the cost per unit area of side S, and the cost per unit area of end E. Then the total cost c to be minimized is

$$c = (VT)l^{-1}h^{-1}w^{-1} + Blw + 2Slh + 2Ehw \equiv t_1 + t_2 + t_3 + t_4$$

Duffin used the values $V = 400$, $T = 0.10$, $B = S = 10$, and $E = 20$, so that their cost was $40(lhw)^{-1} + 10lw + 20lh + 40hw$.

This posynomial has four terms in three variables, for $4 - (3+1) = 0$ degrees of difficulty. Thus the three orthogonality conditions and the single normality condition will have a unique solution. As defined in the preceding section, the exponents α_{tn} are, if the design variables are numbered $x_1 \equiv l$, $x_2 \equiv h$, and $x_3 \equiv w$, as follows: $\alpha_{11} = \alpha_{12} = \alpha_{13} = -1$, $\alpha_{21} = \alpha_{23} = \alpha_{31} = \alpha_{32} = \alpha_{42} = \alpha_{43} = 1$, with all the rest zero. These are the coefficients of the corresponding dual variables in the orthogonality conditions, which are therefore

$$l: \quad -\delta_1 + \delta_2 + \delta_3 = 0$$
$$h: \quad -\delta_1 + \delta_2 + \delta_4 = 0$$
$$w: \quad -\delta_1 + \delta_3 + \delta_4 = 0$$

The normality condition is $\delta_1 + \delta_2 + \delta_3 + \delta_4 = 1$, so the solution is $\delta_1^\star = \frac{2}{5}$; $\delta_2^\star = \delta_3^\star = \delta_4^\star = \frac{1}{5}$. The lower bound is therefore obtained immediately as

$$c \geq \left(\frac{VT}{\frac{2}{5}}\right)^{2/5} \left(\frac{B}{\frac{1}{5}}\right)^{1/5} \left(\frac{2S}{\frac{1}{5}}\right)^{1/5} \left(\frac{2E}{\frac{1}{5}}\right)^{1/5} = 5(V^2T^2BSE)^{1/5}$$

The Duffin values give exactly 100 for this. Since the dual values are the only ones satisfying the normality and orthogonality conditions, this constant must be the *greatest* lower bound, which makes it the minimum value of the objective. All this without ever having even guessed at a design.

Before finding formulas for the design variables, notice that the distribution of the cost at the optimum, being determined by the exponents and not by the parameters, remains constant even when the gravel volume and cost coefficients change, provided the box can be redesigned as needed. Whether the box is made of wood, steel, or, for that matter, hammered gold, the *optimal* design will expend 40% of the total cost on transportation and the rest on the box, equally divided between bottom, sides, and ends. This fixed distribution at the optimum is an example of what will be called here a *design rule*, because with such a rule and the value of the minimum cost, the designer can easily compute the optimal values of the design variables.

To do this, simply use the invariance conditions, which in the example are

$$\frac{VT}{lhw\left(\frac{2}{5}\right)} = \frac{Blw}{\frac{1}{5}} = \frac{2Slh}{\frac{1}{5}} = \frac{2Ehw}{\frac{1}{5}}$$

The common value must of course be the minimum $c_\star = 5(V^2T^2BSE)^{1/5}$, which can be verified by direct substitution into either the objective or dual functions. Multiplication of the first $[5VT/2lhw = 5(V^2T^2BSE)^{1/5}]$ by the second $[5Blw = 5(V^2T^2BSE)^{1/5}]$ gives the height: $h = \frac{1}{2}(VTB^3/S^2E^2)^{1/5}$, which is $\frac{1}{2}$ in the example. Similar manipulation gives $w = (VTS^3/B^2E^2)^{1/5}$ and $l = (VTE^3/B^2S^2)^{1/5}$, respectively 1 and 2 in the example. This expressibility of the entire optimal design in closed form as functions of the parameters is another important characteristic of problems with zero degrees of difficulty.

Be on the lookout for meaningful problems with zero degrees of difficulty. Although they are rare in practice, the discovery of any is an important event. The next section describes one of special interest to mechanical and industrial engineers.

3.12. ZERO DEGREES OF DIFFICULTY: TOLERANCE SPECIFICATION

When there is only one more term than variable, there are no degrees of difficulty, and the normality and orthogonality conditions have an easily obtained unique solution for the dual variables, which therefore must be at their optimizing values $\delta^\star$. These unique values of the dual variables $\delta^\star$, and equivalently the dual weights $\omega^\star$, give elegant design rules of the kind derived for the cofferdam and gravel box problems. Then the dual function $d^\star \equiv d(\delta^\star)$ is not only a lower bound on $p(\mathbf{x})$, it is the *only* and hence the *greatest* lower bound, which makes it the minimum value $p(\mathbf{x}_\star) \equiv y_\star$. Remarkably, this minimum $y_\star$ is known before the corresponding optimal design $\mathbf{x}_\star$, which can be computed later in closed form from the T invariance conditions

$$C_t \prod_{n=1}^{N} x_{n\star}^{\alpha_{ktn}} = \delta_t^\star d^\star$$

Although these are nonlinear, each involves only one variable term, so taking logarithms gives T linear equations in the logarithms of the N unknowns x_n. This is one equation more than needed, but since they must be consistent at the optimum, one need only work with the first $T-1$ terms, using the last as a check. The linear equations are

$$\sum_{n=1}^{N} \alpha_{tn} \log x_{n\star} = \log\left(\frac{\delta_t^\star d^\star}{C_t}\right)$$

The values of $x_{n\star}$ are obtained by exponentiating $\log x_{n\star}$.

More important than the computational advantages of the zero-degree-of-difficulty case is the opportunity to develop an insightful design theory for the problem. Physically and economically meaningful groupings of design variables are suggested by the results, all of which are in closed form. This section develops an example of an important engineering problem which, having zero degrees of difficulty, lends itself to strikingly powerful solution when formulated as a geometric program.

A common problem of the machine designer is to select tolerances z_n for the N parts of an assembly. The total tolerance A for the assembly is fixed by such design considerations as noise, wear, interchangeability, maintenance, and performance. In what is called the "sure-fit" version of this problem, the sum of the tolerances z_n must not exceed the design parameter A, which gives the single constraint

$$\sum_{n=1}^{N} z_n \leq A$$

These tolerances z_n must be nonnegative

$$z_n \geq 0 \qquad n=1,\dots,N$$

In practice, very small tolerances may not be achievable, a situation handled in this model by letting z_n represent the tolerance exceeding some fixed minimal value z_n^0. The actual tolerance is then obtained by adding z_n^0 to the optimized value of z_n. This amounts to a coordinate shift.

The problem is to find the feasible tolerance assignment minimizing the total variable cost of making the parts. The cost of each part decreases with tolerance, so monotonicity analysis on any variable proves that the constraint must be active at the minimum (Fig. 3-2).

$$\sum_{n=1}^{N} z_n \leqq A$$

As z_n becomes very large, the cost asymptotically approaches a minimum value C_n, and let $C_n + R_n$ be the cost for the tightest possible specification,

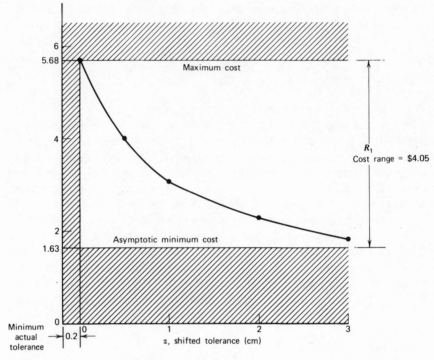

Figure 3-2 Tolerance cost curve for part 1

namely, $z_n = 0$. The difference R_n is called the *cost range* for part n. For intermediate tolerances, each cost is written $C_n + v_n$, with v_n being the *variable cost* for part n found in practice to vary exponentially with z_n (Speckhart).

$$v_n = R_n \exp(-z_n/\tau_n)$$

Here τ_n is a known constant called the *characteristic tolerance* for part n. It is determined empirically either by nonlinear least-squares curve fitting or as the negative reciprocal slope of the straight line obtained by plotting $\ln v_n$ vs z_n, since $\ln v_n = \ln R_n - z_n/\tau_n$. It is convenient to use the same length units for τ_n as for z_n (Fig. 3-3).

As an example, consider an assembly with four parts having dollar costs

$$C_1 + v_1 = 1.36 + 4.05 \exp\left(-\frac{z_1}{0.950}\right)$$

$$C_2 + v_2 = 0.73 + 1.44 \exp\left(-\frac{z_2}{1.215}\right)$$

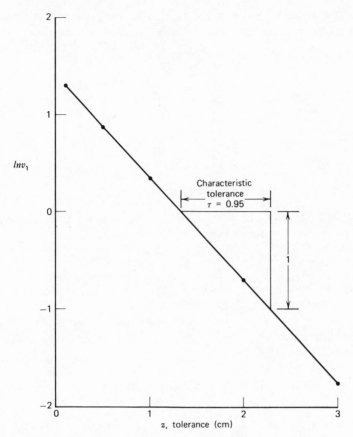

Figure 3-3 Characteristic tolerance determination for part 1

$$C_3 + v_3 = 0.53 + 2.70 \exp\left(-\frac{z_3}{0.662}\right)$$

$$C_4 + v_4 = 1.03 + 3.22 \exp\left(-\frac{z_4}{0.888}\right)$$

The unknown tolerances in centimeters are z_1, z_2, z_3, and z_4; the characteristic tolerances, 0.950, 1.215, 0.662, and 0.888 cm. respectively, and the cost ranges are \$4.05, \$1.44, \$2.70, and \$3.22 respectively.

The problem is to find the tolerances minimizing the total variable cost $v \equiv \Sigma v_n$ subject to the constraint on the total, which in the example takes the form $z_1 + z_2 + z_3 + z_4 \leqq A(\equiv 2.85)$, where the assembly tolerance A is 2.85 cm.

The cost function is not in the right form for geometric programming, since each term is an exponential rather than a power function.

$$v(\mathbf{z}) = \sum R_n \exp\left(-\frac{z_n}{\tau_n}\right)$$

A change of variable $x_n \equiv \exp(-z_n) > 0$ takes care of this problem, giving

$$v(\mathbf{x}) = \sum R_n x_n^{1/\tau_n}$$

The constraint becomes

$$\sum z_n = -\sum \ln x_n = \ln\left(\prod x_n^{-1}\right) \leqq A$$

Taking antilogarithms of both sides and rearranging gives a normalized single term power function inequality.

$$e^{-A} \prod x_n^{-1} \leqq 1$$

Now the problem not only has posynomial form, but also has zero degrees of difficulty, there being $N+1$ terms in N variables.

To solve it, first write the N orthogonality conditions. The constraint requires introduction of the variable δ_{11} for reasons to be explained in Sections 4.01 and 4.02.

$$x_n: \quad \left(\frac{1}{\tau_n}\right)\delta_{0n} - \delta_{11} = 0; \quad n = 1,\dots,N$$

Hence $\delta_{0n}^\star = \tau_n \delta_{11}^\star$ for all n. The normality condition is $\sum \delta_{0t} = 1$.

Substituting the solution of the orthogonality conditions gives

$$\sum \delta_{0n} = \sum \tau_n \delta_{11}^\star = \delta_{11}^\star \sum \tau_n = 1$$

For abbreviation let $\overline{T} \equiv \sum \tau_n$, so that the optimal dual weights $\omega_{0n}^\star$, which equal the optimal dual variables $\delta_{0n}^\star$, are

$$\omega_{0n}^\star = \delta_{0n}^\star = \frac{\tau_n}{\overline{T}}$$

This is interpreted as an elegant design rule:

The optimal variable cost of a part is proportional to its characteristic tolerance.

Notice that the cost ranges R_n need not be known, only the characteristic tolerances τ_n. This phenomenon, in which the optimal proportions are determined entirely by the exponents and not at all by the cost coefficients, always occurs when there are zero degrees of difficulty, for then the optimal weights depend only on the orthogonality and normality conditions generated by the exponents. Using this design rule, an engineer can check a tolerance proposal for optimality merely by inspecting the cost distribution. It is optimal if and only if it matches the characteristic tolerance apportionment. In the example, $\bar{T} = 3.715$, and so $\omega^* = (0.256, 0.327, 0.178, 0.239)^T$.

Now that the optimal weights are known, the dual function, which must equal the minimum variable cost when there are zero degrees of difficulty, can be computed.

$$v_\star = d^\star = \left[\prod \left(\frac{R_n}{\delta_{0n}^\star} \right)^{\delta_{0n}^\star} \right] (e^{-A})^{\delta_{11}^\star} = \bar{T} \prod \left(\frac{R_n}{\tau_n} \right)^{\tau_n / \bar{T}} \left(e^{-A/\bar{T}} \right)$$

This is more readily understood by defining *geometric mean cost range* $\bar{R}$ and *geometric mean characteristic tolerance* $\bar{\tau}$.

$$\bar{R} \equiv \prod (R_n)^{\tau_n / \bar{T}} \qquad \bar{\tau} \equiv \prod (\tau_n)^{\tau_n / \bar{T}}$$

In the example, $\bar{R} = (4.05)^{0.256}(1.44)^{0.327}(2.70)^{0.178}(3.22)^{0.239} = 2.54$, and $\bar{\tau} = (0.950)^{0.256}(1.215)^{0.327}(0.662)^{0.178}(0.888)^{0.239} = 0.950$. Then

$$v_\star = \left(\frac{\bar{R}\bar{T}}{\bar{\tau}} \right) \exp\left(-\frac{A}{\bar{T}} \right)$$

In the example

$$v_\star = \frac{(2.54)(3.715)}{0.950} \exp \frac{-2.85}{3.715} = 4.61$$

It is interesting that the cost ranges and tolerances for the assembly should be averaged geometrically instead of arithmetically. The minimum cost increases directly with geometric mean cost range and assembly characteristic tolerance; inversely with the geometric mean tolerance. Moreover, cost decreases exponentially with the ratio of assembly tolerance to total characteristic tolerance. Two dimensionless ratios emerge as significant: $\bar{T}/\bar{\tau}$ and $\bar{T}/A$; cost increases with both of them. Geometric averaging, an impractically tedious job with logarithms or slide rule, is the work of a minute on a pocket calculator.

With the minimum possible cost known, the designer can evaluate a preliminary proposal to see if the theoretical optimum is worth computing. For example, suppose a set of tolerances $z = (0.75, 0.90, 0.50, 0.70)^T$ is suggested in the illustration. The variable cost is then calculated as 5.26, significantly larger than the theoretical minimum of 4.61. The next task then is to find the theoretically optimal tolerances.

Application of the optimal weight definitions, i.e., the invariance conditions, to the known minimum cost gives

$$
w_{n\star} = \frac{\tau_n}{\overline{T}} = \frac{v_{n\star}}{v_\star} = \frac{R_n \exp(-z_{n\star}/\tau_n)}{(\overline{RT}/\overline{\tau})\exp(-A/\overline{T})}
$$

Taking natural logarithms and solving for $z_{n\star}$ gives

$$
z_{n\star} = \tau_n \left[\frac{A}{\overline{T}} + \ln \frac{R_n \overline{\tau}}{\overline{R}\tau_n} \right] \qquad n = 1, \ldots, N
$$

In the example

$$
z_{1\star} = 0.950 \left[\frac{2.85}{3.715} + \ln \frac{(4.05)(0.950)}{(2.54)(0.95)} \right] = 1.17
$$

The reader can verify that the optimal tolerance schedule is $z_\star = (1.17, -0.06, 0.79, 0.95)^T$, which, however, is infeasible because $z_{2\star}$ is negative.

In the geometric programming analysis, the nonnegativity constraint on the original tolerances was ignored, so negative tolerances can occur as in the example. A quick remedy is to set the offending tolerance to zero and redistribute the deficiency among the others, roughly in proportion to their optimized values. In the example, under the assumption that shop standards require specification no closer than 0.05 cm., a reasonable tolerance schedule would be $(1.15, 0, 0.75, 0.95)^T$ for which the variable cost is 4.62, only 1¢ more than the theoretical, but infeasible, minimum. This would certainly be acceptable in the example, but as an Exercise (3-10), the reader may wish to solve the problem again with the second tolerance set to zero in advance. The problem remaining then has only three variables. This is the procedure to be followed any time an infeasible design cannot be made satisfactory by setting negative variables to zero.

This example has demonstrated why zero-degree-of-difficulty problems are so interesting. Their closed-form solutions give the sort of design formula that engineers can evaluate quickly on a pocket calculator but that totally mystify the layman. In fact, many simple and ancient design formulas can be shown to be derivable as solutions to zero-degree-of-difficulty geometric programs. Once having recognized the power of geometric programming when there are no degrees of difficulty, the designer is ever after on the alert for posynomial optimization problems with exactly one more term than variables.

3.13.   PIPELINE PUMPING STATIONS

Even when a problem has degrees of difficulty that prevent solution in closed form, the optimality conditions generate simple relations between terms that can yield valuable design rules. This concept will be illustrated on the problem of choosing the number of pumping stations on a long pipeline for transporting fluids. The cost objective function, itself of some engineering interest, is developed in this section. Then the next section works out the optimization, gives a simple design procedure, and discusses the general implications for optimization theory and engineering.

Figure 3-4 shows schematically a pipeline intended to convey a known fluid at a given rate between two points at different elevations. The line has a constant slope, and the capital cost for the pumps per unit of power is assumed constant, which implies that the total cost is unaffected by the distribution of power among the stations. The variable annual amortized costs, whose sum is to be minimized, depend only on the number of stations n and the pipe diameter d, which is the same along the whole pipeline.

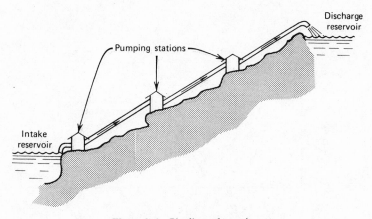

Figure 3-4   Pipeline schematic

The analysis is parametric in the coefficients of the cost terms and in some of the exponents of the variables. Let S be the fixed annual cost of a single station, including pumphouse structure and operating labor, so that the total station cost is Sn. The variable amortized capital cost of the pumps, as well as the cost of the energy needed to overcome friction, are both proportional to $d^{-\alpha}$, where α is a known exponent available from elementary fluid mechanics. In the example in which water is the fluid, $\alpha = 4.87$ from the Hazen-Williams formula. Let F be the proportionality constant so that the friction cost is $Fd^{-\alpha}$.

Amortized pipe cost, including trenching, roads, and crossings is proportional to the weight of the pipe raised to some power β usually slightly less than unity to account for economies of scale. In the example this exponent, which economists call the *elasticity*, is taken as 0.9. It is the value of the logarithmic derivative of the pipe cost wrt the weight, and it can be interpreted as meaning that a 10% increase in weight only produces a 9% increase in cost. Pipe weight is proportional to diameter d and to wall thickness, which for a fixed pressure is also proportional to diameter. Thus pipe cost is proportional to $(d^2)^{\beta} = d^{2\beta}$. The coefficient of proportionality is, however, not a constant.

Instead it depends on the number of stations n. The reason is that a small number of stations means increased elevation difference between stations, as well as increased pressure drop due to friction. Hence the pipeline pressure is higher at the pump, and greater pipe wall thickness is required. Minimum thickness is needed at a pump suction; maximum is required at its discharge. The pipe thickness therefore decreases in the direction of flow between two stations, making the weight dependent on the total number of stations (Fig. 3-5). It can be shown that the cost

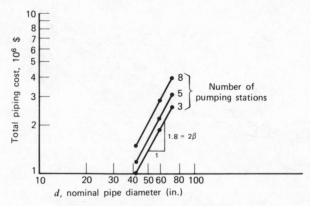

Figure 3-5   Elasticity of total piping cost

coefficient is $(B + En^{-\gamma})$, where B and E are constants and under ideal circumstances $\gamma = 1$. The piping cost is therefore given as the sum of two terms: $Bd^{2\beta} + Ed^{2\beta}n^{-\gamma}$. For reference, the first is called the *basic pipe cost*; the second is called the *extra pipe cost*. The total variable cost is therefore

$$v(n,d) = Sn + Fd^{-\alpha} + Bd^{2\beta} + Ed^{2\beta}n^{-\gamma}$$

3.14. DESIGN RULES DESPITE DEGREES OF DIFFICULTY

The pipeline pumping station problem, with four terms for only two variables, has one degree of difficulty. Yet for each variable there is an interesting design rule, and combination of the two gives the optimal design. The only difference between this and a case with zero degrees of difficulty is that the present one requires a nonlinear computation, whereas all computations are linear when there are no degrees of difficulty.

Let the four terms be given identifying subscripts having the same letters as the cost coefficients: t_s, t_f, t_b, and t_e respectively. Then the optimality condition for the number of stations n is $t_s = \gamma t_e$. The corresponding design rule is

$$\frac{t_s}{t_e} = \gamma$$

In words the ratio of station cost to extra pipe cost at the optimum should be γ. When $\gamma = 1$, as in most cases, the rule is simply that the total station cost should just equal the extra pipe cost.

Another trade-off comes from the optimality condition for the pipe diameter d: $-\alpha t_f + 2\beta t_b + 2\beta t_e = 0$. Rearrangement gives

$$\frac{t_b + t_e}{t_f} = \frac{\alpha}{2\beta}$$

Notice that the two piping cost terms add because d has the same exponent 2β in both. This total is therefore the entire piping cost, the sum of the basic and extra costs. The design rule is therefore that the ratio of piping cost to friction cost should be $\alpha/2\beta$. For water $\alpha = 4.87$, and in the numerical example to follow $\beta = 0.90$, so in that case the pipe cost should be 2.7 times the friction cost.

A design rule establishing a fixed ratio between cost terms can be derived any time a variable has exactly two values for its exponents, as in this pipeline example. These rules lead to simple computations only when there are also zero degrees of difficulty, as in the gravel box problem. Any degrees of difficulty will force the design equations to be nonlinear, which will be seen to be an obstacle easily overcome for pipeline design.

The design rules can be expressed in terms of the design variables to obtain nonlinear equations for the optimum values. The station rule gives

$$Sn = t_s = \gamma t_e = \gamma E d^{2\beta} n^{-\gamma}$$

For a given number of stations n, this can be solved for the corresponding pipe diameter, denoted d_n

$$\frac{d_n}{50} = \left(\frac{Sn^{1+\gamma}}{\gamma E} \right)^{1/2\beta}$$

This is an easily computed increasing function of n. The diameter rule gives

$$Bd^{2\beta} + Ed^{2\beta}n^{-\gamma} = t_b + t_e = \frac{\alpha t_f}{2\beta} = \frac{\alpha F d^{-\alpha}}{2\beta}$$

For given n this can be solved for the corresponding diameter, denoted d_f

$$\frac{d_f}{50} = \left[\frac{\alpha F}{2\beta(B + En^{-\gamma})} \right]^{1/(\alpha + 2\beta)}$$

also an easily computed increasing function of n.

Section 3.10 proved that nonmonotonic posynomials have but one stationary point, and that is a minimum. Thus the problem at hand has a unique stationary point where $d_n = d_f$. For large n, d_n is unbounded above, whereas d_f approaches an asymptote. Hence $d_n > d_f$ for $n > n_\star$, since there is only one $n_\star > 0$ where $d_n = d_f$. This gives a simple, rapidly convergent bracketing scheme for locating $n_\star$ and $d_\star$.

For illustration suppose the cost function has been modeled as

$$v\left(n, \frac{d}{50}\right) = 10^6 \left[0.15n + 3.00\left(\frac{d}{50}\right)^{-4.87} + 1.00\left(\frac{d}{50}\right)^{1.8} + 3.00\left(\frac{d}{50}\right)^{1.8} n^{-1} \right]$$

The number of stations n must of course be an integer. Taking $d = 50$ to make a quick check, one finds the station cost to be $0.150n$; the extra pipe cost is $3.00n^{-1}$. These are equal, as required by the station design rule, when $n = (3/0.150)^{1/2} = (20)^{1/2}$, between four and five. With four stations the total pipe cost is $1.00 + 3.00(4)^{-1} = 1.75$, and the friction cost is 3.00. But their optimal ratio $(\alpha/2\beta = 4.87/1.8 = 2.7)$ is much larger than $1.75/3$, so the pipe, and therefore n, is too small. Thus try $n = 5$ stations first.

For $n = 5$ the station and friction diameters are

$$d_n(5) = 50 \left[\frac{0.150(5^2)}{3.00} \right]^{1/1.8} = 56.5$$

$$d_f(5) = 50 \left[\frac{4.87(3.00)}{1.8(1.00 + 3.00(5)^{-1})} \right]^{1/6.67} = 63.7$$

So $d_n(5) < d_f(5)$, which implies that $n_\star > 5$. Next try $n = 6$.

$$d_n(6) = 69.3 > 64.4 = d_f(6)$$

This bounds $n_\star$ above, so $5 < n_\star < 6$. Despite a degree of difficulty which made the optimal weight distribution depend on the cost parameters, the problem was easily solved numerically. This was because each variable appeared in only a few terms, a characteristic of many engineering design problems.

The bracketing of $n_\star$ between five and six may seem bizarre, for of course n must be an integer. Thus $n_\star$ is fictional in the sense that it violates the requirement that it be a whole number. The next section shows how to find the minimum cost design satisfying the integer requirement. Presently all that can be said is that the minimum cost design will have either five or six stations.

3.15. ROUNDING DISCRETE VARIABLES

Actually, the number of stations is not the only variable required to be a whole number. In practice, large-diameter pipes are only available in certain standard sizes. In the range of interest here, the pipe diameter must be a multiple of 3 in. Both n and d then are examples of *discrete variables*.

If n is fixed, say at an integral value, then $d(n)$ is the diameter minimizing the cost, it having been derived by setting the derivative wrt d to zero. The foregoing computations proved that when $n = 5$, $d_f(5) = 63.7$, and when $n = 6$, $d_f(6) = 64.4$. Since, for fixed n, the cost increases with $|d - d_f(n)|$, the minimum cost for d a multiple of three must be one of the standard sizes bracketing $d_f(5)$ and $d_f(6)$. In both cases these are the diameters 63 and 66. The discrete minimum cost must therefore be the least among the costs of the four cases (n, d): (5,63), (5,66), (6,63), and (6,66), which respectively are 4.15, 4.16, 4.15, and 4.15 million. Since three are the same and the fourth is not much different, there is some flexibility to accomodate ecological, social, or political factors not appearing in the model explicitly (see Fig. 3-6).

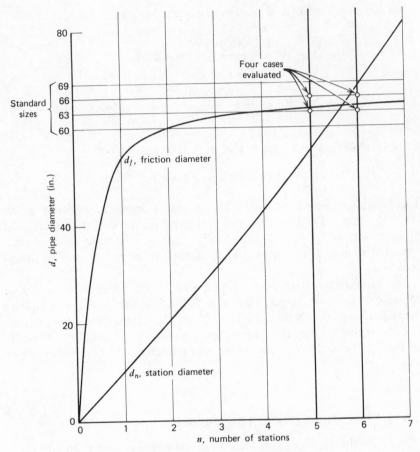

Figure 3-6 Optimal standard design

This concludes the discussion of constructing lower bounds and design rules by geometric programming. In later chapters, these ideas are extended, partially at least, to functions more general than posynomials, still using the powerful geometric inequality.

3.16. DEGREES OF DIFFICULTY: CONDENSATION

Most realistic design problems have at least one degree of difficulty. Three approaches to such problems are presented in the remainder of this chapter. The simplest, called *condensation* and already used on the coffer-dam problem, is demonstrated in this section. A second, more rigorous but slightly more laborious, scheme called *partial invariance* is developed in the next section. Finally the existing computer-oriented dual geometric programming procedure is sketched briefly in Sec. 3.18.

Parametric analysis can be hard and is at best inelegant, when there are degrees of difficulty, so a numerical rather than parametric example will be used. Except for a change in one number, this problem is also from Duffin. He added lengthwise runners to the gravel box of Sec. 3.11, making it into what will be called here for reference a gravel *sled*. To account for the runner cost of 10 per unit length, a fifth cost term $t_5 \equiv 10l$ must be added to the objective, giving

$$c' \equiv c + 10l = 40(lhw)^{-1} + 10lw + 20lh + 40hw + 10l$$

A lower bound on this cost is 100, the cost of the optimal gravel box with no runners. That design $(l, h, w) = (2, \frac{1}{2}, 1)$ would cost 120, the gap of 20 being due to the runners omitted from the earlier optimization. This is too large to tolerate, but before seeking a better design or lower bound, let us choose a satisfaction threshold. The cost coefficients being significant to no more than two figures, a reasonable threshold would be $\tau = 0.01$, which amounts to about one cost unit. On a manual calculator this is handled neatly by not carrying the cost computations past the decimal point. Then a satisfactory design will be any whose cost is indistinguishible from a lower bound when both are rounded to the nearest integer.

The primal weights w_t for the current design are the values of the terms divided by the total, so $w = (\frac{1}{3}, \frac{1}{6}, \frac{1}{6}, \frac{1}{6}, \frac{1}{6})$. The method of *condensation* is to use some of these weights to condense enough terms, two in the example, to obtain a new problem with zero degrees of difficulty. For reasons that may at present seem mysterious, the second and fifth terms will be condensed using equal weights as in the starting design.

$$t_2 + t_5 \geq \left(\frac{10lw}{\frac{1}{2}}\right)^{1/2} \left(\frac{10l}{\frac{1}{2}}\right)^{1/2} = 20lw^{1/2}$$

Then c' is bounded below by the following posynomial with zero degrees of difficulty:

$$c' \geq 40(lhw)^{-1} + 20lw^{1/2} + 20lh + 40hw$$

The technique is called condensation because the objective is condensed *before* computing the lower bound. On the other hand, the partial invariance method of the next section delays condensation until *after* the lower bound is calculated.

This objective is simple enough that the lower bound can be computed without writing down the orthogonality and normality conditions by a process resembling dimensional analysis. Simply write the dual function with the primal variables included and with spaces left for the dual variable exponents and denominators.

$$c' \geq \left[\left(\frac{40}{lhw} \right) \left(\frac{20lw^{1/2}}{} \right) \left(\frac{20lh}{} \right) \left(\frac{40hw}{} \right) \right]$$

Then choose exponents that make the product dimensionless. For instance; for the numerators to have the variables all raised to the same power, the exponent on the third factor must be half that on the second and fourth. Taking unity for this exponent, the exponents must be 2 on the second and fourth, which forces an exponent of 3 on the first factor. This takes care of orthogonality; to make them sum to unity as required by the normality condition, simply raise the entire function to the power $[3+2+1 +2]^{-1}$. The dual variables, obtained by multiplying exponents, can be written in the denominators as shown.

$$c' \geq \left\{ \left[\frac{40}{3/8\,lhw} \right]^3 \left(\frac{20lw^{1/2}}{2/8} \right)^2 \left(\frac{20lh}{1/8} \right)^1 \left(\frac{40hw}{2/8} \right)^2 \right\}^{1/8}$$

If this is too confusing at first, just solve the orthogonality and normality conditions in the usual way to get the same answer. The lower bound turns out to be 116.

Since the gap of four exceeds the satisfaction threshold, a new design is generated from the invariance conditions for this zero-degree-of-difficulty problem. Finding this design is left at Exercise 3-14. Most important, the associated cost is 116, indistinguishible from the lower bound. *Satis quod sufficit*—enough is as good as a feast.

"Impressive," you might say, "but how did you decide on the second and fifth terms for this magical condensation?" The answer is to try to condense terms whose exponents are not very different. On physical grounds this is equivalent to constructing an approximation as nearly like the original as possible. Especially avoid condensing terms for which the exponents for some variable have opposite signs, for this immediately generates a monotonic approximation to a nonmonotonic function. The trick is to keep away from condensations that make the lower-bounding function unbounded. In large problems with constraints, it may not be easy to see how to do this, in which case the strategy of the next section is in order.

3.17. PARTIAL INVARIANCE

The primal weights for a starting design were used in condensation to construct a lbf with zero degrees of difficulty. They can also be used to compute a lower bound directly without condensation. To do this, designate a number of terms, equal to the number of design variables plus one, as *dominant*. These should be the ones expected to be the largest in the optimal design. In the example these would be the first four, since the fifth was regarded as negligible in generating the starting design.

Next write down the orthogonality and normality conditions. In the gravel sled example these are:

$$
\begin{array}{ll}
l: & -\delta_1 + \delta_2 + \delta_3 \qquad + \delta_5 = 0 \\
h: & -\delta_1 \qquad + \delta_3 + \delta_4 \qquad = 0 \\
w: & -\delta_1 + \delta_2 \qquad + \delta_4 \qquad = 0 \\
& \delta_1 + \delta_2 + \delta_3 + \delta_4 + \delta_5 = 1
\end{array}
$$

Solve these linear equations for the *dominant* dual variables, that is, the ones corresponding to the dominant terms. In the example this gives the four dominant duals as linear functions of the *recessive* δ_5.

$$
\delta_1 = \frac{2 - \delta_5}{5} \qquad \delta_2 = \frac{1 - 3\delta_5}{5} \qquad \delta_3 = \frac{1 - 3\delta_5}{5} \qquad \delta_4 = \frac{1 + 2\delta_5}{5}
$$

To keep all of them positive, δ_5 must be less than $\frac{1}{3}$. Since the primal weight w_5 is $\frac{1}{6}$, right in the middle of this range, let $\delta_5 = \frac{1}{6}$, which generates the set of duals $\delta = (11, 3, 3, 8, 5)/30$. The reader can verify that the dual function for these values takes the value 113.2, where the figure after the decimal point is not significant, being carried to reduce roundoff error.

This is seven units less than the starting design cost of 120, so a new design must be sought, this time from the invariance conditions. The problem is that they will not be consistent unless by chance the lower bound is at the exact optimum. Thus any set of three invariance conditions will give values of the three design variables, but these values will not satisfy the other two conditions. Which three would be reasonable to use?

The three largest dual variables are δ_1, δ_4, and δ_5, for which the invariance conditions are $40(lhw)^{-1} = (113.2)11/30$; $40hw = (113.2)8/30$; and $10l = (113.2)5/30$. Unfortunately, this subset of invariance conditions is internally inconsistent, as the reader can verify by multiplying the three equations together. This shows that not every collection of three invariance conditions will do—they must also be consistent.

Let us therefore replace δ_5, the smallest of these three, by one of the others, δ_2 or δ_3, which both have the same value. Arbitrarily choose the latter, whose invariance condition is $20lh = (113.2)3/30$. The set is now consistent, giving the design $(l, h, w) = (1.28, 0.44, 1.70)$. The cost associated is 117.5, leaving an unacceptable gap of four units. Iteration of this procedure, using the new distribution of costs, is left as Exercise 3-15.

This procedure for generating a design from a dual solution is called the *partial invariance* method. In this example it did not perform as well as condensation, but there are situations where the latter cannot be applied at all. One is when no starting solution is available to generate the weights for condensation. The other, encountered in the next chapter, happens when there are constraints to consider. In these cases partial invariance can still be used, although condensation cannot.

3.18.   DUAL GEOMETRIC PROGRAMMING

When there is but one degree of difficulty, it is easy to evaluate the dual function at several values of the single recessive dual variable. Thus one can search directly for the value giving the *greatest* lower bound, which is by definition the value of the minimized objective function. This procedure will first be demonstrated on the example and then justified theoretically.

Let $d(\delta_5)$ be the dual function $d[\boldsymbol{\delta}(\delta_5)]$, it being understood that the dominant duals are determined by the *recessive* (i.e., not dominant) δ_5 to satisfy the orthogonality and normality conditions. This dual function is known now at $\delta_5 = 0$, where $d(0) = 100$, and at $\delta_5 = \frac{1}{6}$, where $d(0.167) = 113.2$. In the middle of this interval, it is a matter of minutes on a pocket calculator to compute $d(0.0833) = 115.0$, which verifies that the maximizing value δ_5^* is in the interval $0 < \delta_5^* < 0.167$. Since $d(0) < d(0.167)$, δ_5^* is more likely to be in the interval $(0.0833, 1.67)$ than in $(0, 0.0833)$, and at the midpoint of the former, $d(0.125) = 115.6$. Now the interval $(0.0833, 0.125)$ looks attractive, and at its midpoint, $d(0.104) = 115.7$.

This procedure could be continued indefinitely, or at least until an acceptably small range containing δ_5^* is found. But since the dual function varies less than a unit in the last interval, one could hazard evaluating the objective to obtain an upper bound on the dual function being maximized. The exact optimum has perhaps not been achieved, but the partial invariance method can always be attempted. Since the present point is more nearly optimal than that of the preceding section, it certainly has a better chance of success. The corresponding design, $(l, h, w) = (1.289, 0.589, 1.207)$ gives $c' = 115.7$, indistinguishable to the first decimal place from the dual solution lower bound. Hence the search can terminate after three iterations.

To understand how this approach extends to any number of dual variables, recall the inequality of Sec. 3.10 relating the objective and dual functions.

$$p(\mathbf{x}) \equiv \sum_{t=1}^{T} C_t \prod_{n=1}^{N} x^{\alpha_{tn}} \geq \prod_{t=1}^{T} \left(\frac{C_t}{\delta_t} \right)^{\delta_t} \equiv d(\boldsymbol{\delta})$$

where $C_t > 0$, α_{tn} are real, and the components of $\boldsymbol{\delta}$ satisfy $\boldsymbol{\delta} \geq \mathbf{0}$, the N orthogonality conditions

$$\sum_{t=1}^{T} \alpha_{tn} \delta_t = 0 \ (n = 1, ..., N)$$

and the normality condition

$$\sum_{t=1}^{T} \delta_t = 1$$

Minimizing $p(\mathbf{x})$ is called the *primal geometric programming problem*, or *primal* for short. The minimum, written $p_\star$ or $\min_x p$, is by definition the *greatest* lower bound on $p(\mathbf{x})$. Since any nonnegative $\boldsymbol{\delta}$ satisfying the orthogonality and normality conditions gives a dual function $d(\boldsymbol{\delta})$ which is a lower bound on $p(\mathbf{x})$, the greatest lower bound $p_\star$ is the same as the maximum among all possible $d(\boldsymbol{\delta})$. Hence it can be found by maximizing $d(\boldsymbol{\delta})$ wrt $\boldsymbol{\delta}$, subject to nonnegativity, orthogonality, and normality. This is called the *dual geometric programming problem*

$$\max_{\boldsymbol{\delta} \geq 0} \prod_{t=1}^{T} \left(\frac{C_t}{\delta_t} \right)^{\delta_t}$$

subject to

$$\sum_{t=1}^{T} \alpha_{tn} \delta_t = 0 \qquad n = 1, ..., N \qquad \text{orthogonality}$$

$$\sum_{t=1}^{T} \delta_t = 1 \qquad \text{normality}$$

The values $\boldsymbol{\delta}^\star$ maximizing the dual function $d(\boldsymbol{\delta})$ generate the values $\mathbf{x}_\star$ minimizing the primal function through the invariance conditions.

$$C_t \prod_{n=1}^{N} \frac{x_n^{\alpha_{tn}}}{\delta_t^\star} = d(\boldsymbol{\delta}^\star) = p_\star \qquad t = 1, ..., T$$

Taking common logarithms gives T equations linear in the N variables $\log x_n$; these are consistent at the optimum and are convenient for numerical computation when the invariance conditions are strongly coupled.

$$\sum_{n=1}^{N} \alpha_{tn} \log x_n = \log\left(\frac{\delta_t^{\star} p_{\star}}{C_t} \right) \qquad t = 1, \ldots, T$$

When as in the example there is only one degree of difficulty, fixing the value of a single dual variable determines all the others, provided the orthogonality and normality conditions are linearly independent. In addition, the value chosen must not only be nonnegative itself, it has to generate nonnegative values of the others. Then the dual function can be evaluated quickly. Finding the maximum value of the dual function is easily accomplished by a direct search of the limited range of allowable values of the recessive dual variable.

This search can be accelerated by exploiting the fact that the only stationary point in the interval is the maximum sought, a property known to optimization theorists as *unimodality*. On a unimodal function, the location of the maximum sought must be bracketed by the dual values on each side of that where the largest value of the dual function has been found. Thus the first three guesses narrow the interval, and each one afterward eliminates more. The simple "bisection" strategy used in the example cuts the interval in half each time. The even more effective "Golden Section" procedure is described in optimization theory texts like *Foundations of Optimization*, as well as in Sec. 8.2.

When there are many degrees of difficulty, direct search unaided either by more sophisticated optimization theory or, as in later sections of this book, by engineering experience, is wasteful. The dual problem is then more effectively solved numerically as one of maximizing a nonlinear function subject to linear inequality constraints. Very efficient computer codes based on advanced principles of nonlinear programming are now in most computer centers, that of Dembo being as good as any.

When a computer code is available, dual geometric programming is certainly the best way to solve an unconstrained posynomial minimization problem numerically. Condensation and partial invariance can then be regarded as quick but rigorous ways to check out a designer's hunch without calling in the computer people. There is, however, another reason for knowing these other methods. They will work when some of the terms are negative, a situation violating the premises from which dual geometric programming was derived. This more general case will be studied in Chapters 5 and 6, after posynomial constraints have been discussed.

3.19.   CONCLUDING SUMMARY

This chapter has shown how to make the estimation process rigorous by manipulating inequalities to construct bounds. Much computation and analysis can be saved by seeking satisfactory rather than theoretically optimal designs: "*Satis quod sufficit.*"

Many engineering design problems are seen to involve power functions that, being linear in the logarithms of their variables, have very simple semilogarithmic and logarithmic derivatives. Such problems can be solved by methods related to geometric programming, which yields the useful concept of "degrees of difficulty," teiling the designer how many items must be neglected or estimated in a problem to generate an easily computed bound. Cases with zero degrees of difficulty, as in the gravel box and tolerance assignment examples, are strikingly easy to solve and lend themselves to the formulation of elegant design rules. Even when there are degrees of difficulty, as in the pipeline example, similar design rules can be extracted from the orthogonality conditions.

Three procedures for improving designs—condensation, partial invariance, and dual geometric programming—have been presented. Dual geometric programming is well suited for exact optimization on a computer, but it can only be used on posynomials. The next chapter shows how to extend these procedures to design problems with posynomial constraints. Chapter 5 applies condensation and partial invariance to objectives with negative terms. Partial invariance can be used even when, as in Chapter 6, there are negative terms in the constraints that prevent construction of lower bounds.

There is one modern language, Brazilian Portugese, which expresses most naturally the attitude of this book, especially in this chapter. When, in Rio de Janeiro, one Carioca asks another how things are going, the answer will often be the Roman thumbs up sign and the Portugese exclamation, "*ótimo!*" (pronounced AHcheemoo). Of course, both *ótimo* and the English word "optimal" descend from the Latin *optimo*, meaning "best." But the Carioca certainly does not mean that an extensive study has rigorously determined that his personal objective has achieved the best value possible. By *ótimo* he asserts that he knows how good things can be, and that he is close enough to the best that further effort not only is unjustified, but also would impede his enjoyment of this happy state. The cartoon, a Portugese translation of Charles Schulz' insightful comic strip *Peanuts*, illustrates this attitude well. Like Snoopy the sensible beagle and Zé the casual Carioca, a good designer knows when things are "great."

PEANUTS   *Charles M. Schulz*

Figure 3-7 How was the golf game? (Cartoon used by permission of United Feature Syndicate, Inc.)

NOTES AND REFERENCES

Using inequalities to construct lower bounds is widespread in operations research because of the frequent linearity of scheduling problems. Beckenbach and Bellman gathered together nonlinear inequalities of potential application to engineering. Hardy, Littlewood, and Polya wrote the definitive work on classical inequalities. Eben and Ferron early recognized the possibilities, especially those for finding invariance principles with physical interpretations in the staged operations of chemical engineering. But the strongest case for constructing lower bounds in design was made by Duffin in his 1962 article defining geometric programming.

Zener had previously noticed the special character of power functions, especially when there are no degrees of difficulty. His more recent book on design applications of geometric programming abounds with examples, and that of Duffin, Peterson, and Zener is the basic reference on geometric programming. The research-minded designer may wish to study the "equilibrium conditions" given there, for they are not used at all in this book. Much of the recent work on formal solution algorithms for geometric programming is surveyed by Beightler and Phillips. The name and technique of condensation come from Duffin's 1970 article.

The tolerance problem started as Ed Prentice's student project in optimal design.

Beckenbach, E., and R. Bellman, *An Introduction to Inequalities*, Random House, New York, 1961.

Beightler and Phillips, *op. cit.* in Chapter 1.

Duffin, R. J., "Cost Minimization Problems Treated by Geometric Means," *Oper. Res.*, **10**, 669 (1962).

Duffin, R. J., "Linearizing Geometric Programs," *Soc. Indust. Appl. Math. Rev.*, **12**, 211–227 (1970).

Duffin, Peterson, and Zener, *op. cit.* Chapter 1.

Eben, C. D., and J. R. Ferron, "Inequality Methods for Computation of Optimal Systems," *Ind. Eng. Chem. Fund.*, **8**, 749–757 (1969).

Hardy, G. H., J. E. Littlewood, and G. Polya, *Inequalities*, Cambridge University Press, Cambridge, England, 1959.

Speckhart, F. M. "Calculation of Tolerance Based on a Minimum Cost Approach," *Trans. ASME, J. Eng. Ind.* 71-Vibr-114 (1971).

Wilde, D. J., and E. Prentice, "Minimum Exponential Cost Allocation of Sure-Fit Tolerances," *Trans. ASME, J. Eng. Ind.* **97**, B4 (Nov 1975) 1395-1398.

Zener, C., "A Mathematical Aid in Optimizing Engineering Design," *Proc. Natl. Acad. Sci.*, **47**, 537 (1961).

Zener, *op. cit.* Chapter 1.

EXERCISES

3-1. Write the first semilog derivatives for the gravel box of Sec. 3.11. Compare them with the orthogonality conditions. Evaluate them for the design $l = h = w = 1$ (Sec. 3.1).

3-2. Write the second semilog derivatives for the gravel box of Sec. 3.11 and evaluate them at $(l, w, h) = (2, 1, \frac{1}{2})$ (Sec. 3.1).

*3-3. Identify all functions in your design project that are posynomials (Sec. 3.2).

*3-4. Identify any terms in your design project that may be negligible (Sec. 3.3).

3-5. Prove by mathematical induction that $\Sigma x_i / n \geq \Pi x_i^{1/n}$ with equality if and only if $x_1 = x_2 = \cdots = x_n$ (Sec. 3.4.).

*3-6. On any posynomial in your design project, construct a lower-bounding function (Sec. 3.5) and a lower-bounding constant (Sec. 3.6).

*3-7. Select a satisfaction threshold for your design project (Sec. 3.7).

3-8. Write the term optimality conditions, the weight optimality conditions, and the orthogonality conditions for the pipeline pumping station problem of Sec. 3.14 (Sec. 3.8).

3-9. Using the parametric results of Sec. 3.11, redesign the gravel box for a pile volume of $V = 800$, twice as large as before. Notice that the cost does not double.

3-10. Solve the tolerance problem of Sec. 3.12 again, with the second tolerance set to zero in advance. Compare this exact result with the approximate one obtained in the text.

*3-11. Identify any economic elasticities in your design project (Sec. 3.13).

*3-12. Can you find any simple design rules in your design project (Sec. 3.14)?

3-13. Double the station cost for the pipeline problem and find the corresponding optimal design for an integral number of stations, and pipe diameters a multiple of 6 in. (Sec. 3.15).

3-14. Find the gravel sled design for the zero degree of difficulty problem at the end of Sec. 3.16.

*Project problem.

3-16. Iterate the partial invariance procedure of Sec. 3.17 one more time.

3-17. Use dual Geometric Programming on the pipeline problem of Sec. 3.14, taking $n = 5$ and $d = 60$ as a base case. Allow a nonintegral number of stations and a nonstandard pipe diameter (Sec. 3.18).

3-18. In W. Braga's fruit van problem (Exercise 2-8), substitution of the constraints gives the following unconstrained cost function: $62(10^7)s^{-3} + 25(10^{-4})s^2 t + 35(10^4)s^{-1}(t + 1.2)^{-1} + 96(10^{-4})s^2$. The constant 1.2 represents the insulating value of the air surrounding the van, expressed as an equivalent thickness of insulation. Neglect this constant as a first approximation, to have the objective in posynomial form. Then find a design whose cost is within 1% of the true minimum. (Hint: construct a base case by neglecting the last term, representing material cost).

CHAPTER 4

Conditional Design
and Constraint Activity

Can one desire too much of a good thing?

Miguel de Cervantes, *Don Quixote* (1610)

After preliminary studies, designers often find they have "too much of a good thing." The design, while technically optimal or at least satisfactory, just doesn't look right—something is too big, too small, too fast, or out of proportion. When this happens, closer examination usually shows that some constraint was overlooked which, when included in the model, replaces the first design with a more reasonable one.

The present chapter deals with inequality constraints that, like the objective function, are posynomials—sums of power functions. This involves straightforward extension of the geometric programming theory of Chapter 3 to the inequality constrained case. For this theory to work efficiently, one must know which inequalities will, at the optimum, be active, i.e., satisfied as strict equalities. The monotonicity analysis of Chapter 2 is useful for this, and other ways of determining constraint activity will be developed.

Although there may be in principle many combinations of tight and loose constraints, actually only a small number can satisfy the conditions required for optimality. These few potentially optimal combinations may in fact each represent a possible optimal design, the values of the design parameters determining which particular combination is optimal in a given situation. When the number of potentially optimal combinations is small, and each is easy to compute because of few degrees, either of freedom or of difficulty, the best design procedure may be to calculate all of them and select the optimum by direct comparison. Usually, however, total enumeration is not needed. One can first evaluate the possibly optimal design having the fewest constraints active. Then if all other constraints turn out to be inactive as predicted, the first design must be optimal. If, on the other hand, any constraints are violated, the possibly optimal combination including these constraints is checked.

When each possibly optimal combination has a closed form solution, this approach can be regarded as a *conditional* design procedure, in which the designer merely evaluates formulas systematically, the results of each step determining which calculation to perform next. Conditional design procedures avoid numerical iteration, give the designer complete control of the calculations, and furnish insight into the types of optimal design possible as the parameters change.

In this chapter the examples range from a simple hydraulic cylinder to a fleet of superships. Both problems can be modeled as geometric programs with posynomial objective and constraints. Although, like the pressure vessel, cofferdam, and pipeline design examples of the earlier chapters, these could be solved numerically for any given parameters by dual geometric programming, much simpler design procedures will be developed. The engineering insights gained are interesting, not only for the noncontroversial hydraulic cylinder, but especially for the supership design, which has been raising the hackles of the ecologically minded all over the world.

4.1. CONSTRAINED POSYNOMIALS

Posynomial minimization problems with posynomial constraints have special properties of advantage to the designer. The main one is that such problems can have at most one stationary point, which must be a global minimum. This follows from the geometric programming construction derived in the next section, which ensures that a constant lower bound can always be computed, provided all constraints are posynomials bounded above by unity. Another advantage is that condensation is always applicable to problems with degrees of difficulty, although some care is needed to avoid unboundedness.

These issues can be illustrated on the cofferdam problem as posed before any variables were eliminated. In geometric programming form, with the values of the parameters given, the primal problem is $\min_{f,h,s,t} p_0(f,h,s,t)$, with

$$p_0 = 168st + 3660t + 3660f^{-1}st + 50,000h^{-1}$$

subject to

$$1.0425s^{-1}t \le 1 \qquad 0.00035ft \le 1 \qquad ht^{-1}+41.5t^{-1} \le 1$$

Notice that the constraints known to be active are written as posynomial functions bounded above by unity, in which form they are said to be *normalized*.

Introduction of a dual weight for each objective term and subsequent application of the geometric inequality gives

$$p_0 = \left(\frac{168st}{\omega_{01}}\right)^{\omega_{01}} \left(\frac{3660t}{\omega_{02}}\right)^{\omega_{02}} \left(\frac{3660f^{-1}st}{\omega_{03}}\right)^{\omega_{03}} \left(\frac{50{,}000h^{-1}}{\omega_{04}}\right)^{\omega_{04}}$$

$$= \left(\frac{168}{\omega_{01}}\right)^{\omega_{01}} \left(\frac{3660}{\omega_{02}}\right)^{\omega_{02}} \left(\frac{3660}{\omega_{03}}\right)^{\omega_{03}} \left(\frac{50000}{\omega_{04}}\right)^{\omega_{04}} s^{\omega_{01}+\omega_{03}} t^{\omega_{01}+\omega_{02}+\omega_{03}} f^{-\omega_{03}} h^{-\omega_{04}}$$

Resist the urge to set the exponents of the four variables to zero to obtain a constant lower bound. This would be incorrect because it would disregard the constraints. Instead one must raise each constraint function to a positive power yet to be determined. Thus for every positive value of the constants λ_1, λ_2, and λ_3 the following inequalities hold:

$$1 \ge (1.0425s^{-1}t)^{\lambda_1} \qquad 1 \ge (0.00035ft)^{\lambda_2} \qquad 1 \ge (ht^{-1}+41.5t^{-1})^{\lambda_3}$$

$$\ge \left[\left(\frac{1}{\omega_{31}}\right)^{\omega_{31}} \left(\frac{41.5}{\omega_{32}}\right)^{\omega_{31}} h^{\omega_{32}} t^{-1}\right]$$

Notice the use of the geometric inequality in the third constraint. Multiplication of p_0 by the left members, all unity, of these inequalities of course gives p_0 again. The sense of the inequalities is preserved by multiplying left members, which gives the single inequality

$$p_0 \ge \left(\frac{168}{\omega_{01}}\right)^{\omega_{01}} \left(\frac{3660}{\omega_{02}}\right)^{\omega_{02}} \left(\frac{3660}{\omega_{03}}\right)^{\omega_{03}} \left(\frac{50000}{\omega_{04}}\right)^{\omega_{04}} \times$$

$$\times (1.0425)^{\lambda_1} (0.00035)^{\lambda_2} \left(\frac{1}{\omega_{31}}\right)^{\lambda_3\omega_{31}} \left(\frac{41.5}{\omega_{32}}\right)^{\lambda_3\omega_{32}} \times$$

$$\times s^{\omega_0+\omega_{03}-\lambda_1} t^{\omega_{01}+\omega_{02}+\omega_{03}+\lambda_1+\lambda_2-\lambda_3} f^{\omega_{03}+\lambda_2} h^{-\omega_{04}+\lambda_3\omega_{31}}$$

There is consequently an orthogonality condition for each of the four variables obtained by setting the exponents to zero. The conditions are:

$$
\begin{array}{llll}
s: & \omega_{01} + \omega_{03} & -\lambda_1 & = 0 \\
t: & \omega_{01} + \omega_{02} + \omega_{03} & +\lambda_1 + \lambda_2 - \lambda_3 & = 0 \\
f: & -\omega_{03} & +\lambda_2 & = 0 \\
h: & -\omega_{04} & +\lambda_3 \omega_{31} = 0 &
\end{array}
$$

In addition, the dual weights for the objective must sum to unity

$$
\omega_{01} + \omega_{02} + \omega_{03} + \omega_{04} = 1
$$

as must those for the third constraint.

$$
\omega_{31} + \omega_{32} = 1
$$

These six equations have nine unknowns, so three variables must be specified before the others can be determined. However, there are only four primal variables, and since all three constraints have been proven active at the optimum, specifying any design variable will fix the other three. Hence a base case will be generated by estimating the value of one design variable and computing the other three. Let us choose $t = 42$. Solution of the constraints gives $s = 43.8$, $f = 68.0$, and $h = 0.5$. Let dimensionless variables be introduced that are unity in this base case.

$$
x_1 \equiv f/68.0 \qquad x_2 \equiv h/0.5 \qquad x_3 \equiv s/43.8 \qquad x_4 \equiv t/42
$$

The cost being $661.8(10^3)$ for these values, the objective is scaled by $y \equiv c/661.8(10^3)$. The resulting scaled problem is

$$
\min y: \quad y \equiv 0.467 x_3 x_4 + 0.232 x_4 + 0.150 x_1^{-1} x_3 x_4 + 0.151 x_2^{-1}
$$

subject to

$$
x_3^{-1} x_4 \leqq 1 \qquad x_1 x_4 \leqq 1 \qquad 0.01190 x_2 x_4^{-1} + 0.988 x_4^{-1} \leqq 1
$$

Notice that the coefficients of the scaled problem are the values of the primal weights at the base case where $\mathbf{x} = \mathbf{1} \equiv (1, 1, 1, 1)^{\mathrm{T}}$.

A lower bound is now constructed by condensing the problem to reduce the degrees of difficulty to zero. There are four ways to do this: the objective can be condensed into a single term, or else any three terms of the objective, together with the two terms of the constraint p_3. Certainly the first three terms of the cost, which concern the construction cost, should be condensed, since the distribution among them is unlikely to change much as the cofferdam height varies. On the other hand, the flood risk cost, being inversely proportional to the height above the water, is too sensitive to be included in a condensation. The remaining degree of difficulty can therefore be removed only by approximating the two terms of p_3 by a single power function.

The base case weights and the geometric inequality give

$$1 \geqq p_3 = 0.0119 x_2 x_4^{-1} + 0.988 x_4 \geq x_2^{0.0119} x_4^{-1} \equiv \bar{p}_3$$

Condensation of the first three terms of the objective gives a lower bounding function.

$$p_0 = 0.849\big(0.550 x_3 x_4 + 0.273 x_4 + 0.177 x_1^{-1} x_3 x_4\big) + 0.151 x_2^{-1}$$
$$\geq 0.849 x_1^{-0.177} x_3^{0.727} x_4 + 0.151 x_2^{-1} \equiv l(\mathbf{x})$$

Introduction of dual weights ω_{01} and ω_{02} for these terms, application of the geometric inequality, and multiplication of the resulting lower bound by the functions $p_1^{\lambda_1}$, $p_2^{\lambda_2}$, and $\bar{p}_3^{\lambda_3}$ yields

$$p_0 \geq l(\mathbf{x}) p_1^{\lambda_1} p_2^{\lambda_2} \bar{p}_3^{-\lambda_3}$$
$$\geq \left(\frac{0.849 x_1^{-0.177} x_3^{0.727} x_4}{\omega_{01}}\right)^{\omega_{01}} \left(\frac{0.151 x_2^{-1}}{\omega_{02}}\right)^{\omega_{02}} \left(\frac{x_4}{x_3}\right)^{\lambda_1} (x_1 x_4)^{\lambda_2} \left(\frac{x_2^{0.0119}}{x_4}\right)^{\lambda_3}$$

This expression is made dimensionless by the unique set of exponents $\omega_{01} = 0.978$, $\omega_{02} = 0.022$, $\lambda_1 = 0.711$, $\lambda_2 = 0.173$, and $\lambda_3 = 1.862$, which give a lower-bounding constant of 0.91. This leaves an unacceptable 9% gap between the base case cost and the lower bound.

The "design" that is the primal solution to the zero degree of difficulty problem which generated the lower bound is $\mathbf{x} = (0.977, 7.48, 1.024, 1.024)$, but it cannot be used because it violates constraint p_3.

$$p_3 = 0.0119(7.48)(1.024)^{-1} + 0.988(1.024)^{-1} = 1.052 > 1$$

This should not be surprising, since $\bar{p}_3$, set to unity to construct the zero-degree-of-difficulty problem, is a lower-bounding function for the original constraint p_3.

A feasible design is easily obtained, however, by changing x_2 to 3.04 to satisfy p_3, while retaining all the other values. Then $p_0(0.977, 3.04, 1.024, 1.024) = 0.94$, which closes the gap to 3%. Iterating this condensation procedure with the weights from the new design is left as Exercise 4-1.

This example had two objectives. One was to show how to use condensation on a posynomial constraint, obtain a valid lower bound, and generate a feasible design from the infeasible one corresponding to the zero degree of difficulty problem. The second, more important, goal was to set the stage for extending the elegant and powerful theory of geometric programming to the case where there are posynomial constraints. The experience gained in working through this example should make the abstract presentation of the next section less difficult to understand.

4.2. GEOMETRIC PROGRAMMING

The preceding example will now be generalized to derive the solution to an important class of engineering problems, namely, those involving only posynomials. Let there be N independent (actually, they are *inter*dependent when there are constraints) positive variables $x_1, \ldots, x_N$, with the column vector $(x_1, \ldots, x_N)^T$ abbreviated as $\mathbf{x}$.

$$\mathbf{x} \equiv (x_1, \ldots, x_N)^T > \mathbf{0} \qquad \text{the zero vector}$$

The objective function, to be minimized wrt $\mathbf{x}$, is written $p_0(\mathbf{x})$. It is a sum of T_0 positive power functions of $\mathbf{x}$, each called a *term* $t_{0t}(\mathbf{x})$, where the subscript t ranges from 1 to T_0 to identify the term.

$$p_0(x) \equiv \sum_{t=1}^{T_0} t_{0t}(\mathbf{x})$$

A typical term $t_{0t}(\mathbf{x})$, a power function of $\mathbf{x}$, has the form

$$t_{0t}(\mathbf{x}) = C_{0t} \prod_{n=1}^{N} x_n^{\alpha_{0tn}}$$

where C_{0t} is a positive parameter, and the exponents α_{0tn} are any real numbers, positive or negative. Three subscripts are needed: the first identifies the exponent as being in the objective; the second identifies the term; the third, the design variable.

Suppose there are K normalized inequality posynomial constraints, all known to be satisfied as strict equalities at the minimum.

$$p_k(\mathbf{x}) \leqq 1 \qquad\qquad k = 1, \ldots, K$$

where

$$p_k(\mathbf{x}) \equiv \sum_{t=1}^{T_k} t_{kt}(\mathbf{x}) \qquad\qquad k = 1, \ldots, K$$

and

$$t_{kt}(\mathbf{x}) \equiv C_{kt} \prod_{n=1}^{N} x_n^{\alpha_{ktn}} \qquad t = 1, \dots, T_k$$

Here T_k is the number of terms in the kth constraint function, and as for the objective function, C_{kt} is a positive parameter and α_{ktn} a real exponent.

The *primal geometric programming problem* is

$$\min_{\mathbf{x} > \mathbf{0}} p_0(\mathbf{x})$$

subject to

$$p_k(\mathbf{x}) \leqq 1 \qquad k = 1, \dots, K$$

This is also called the *posynomial primal problem*.

The solution will be obtained by constructing weighted geometric means of all the posynomial functions. Thus define a positive *dual weight* ω_{kt} corresponding to each term t_{kt}, where now k ranges from 0 through K to include the objective $p_0(\mathbf{x})$. The dual weights must sum to unity for each posynomial.

$$\sum_{t=1}^{T_k} \omega_{kt} = 1 \qquad k = 0, 1, \dots, K$$

These $K + 1$ sums are called *normality conditions*.

Applying the geometric inequality to each posynomial gives

$$p_k(\mathbf{x}) = \sum_{t=1}^{T_k} \omega_{kt} \left(C_{kt} \prod_{n=1}^{N} x_n^{\alpha_{ktn}} / \omega_{kt} \right)$$

$$\geq \prod_{t=1}^{T_k} \left[\frac{C_{kt} \prod_{n=1}^{N} x_n^{\alpha_{ktn}}}{\omega_{kt}} \right]^{\omega_{kt}}$$

$$= \prod_{t=1}^{T_k} \left(\frac{C_{kt}}{\omega_{kt}} \right)^{\omega_{kt}} \prod_{n=1}^{N} x_n^{\alpha_{ktn}\omega_{kt}}$$

$$\equiv d_k(\omega) \prod_{n=1}^{N} x_n^{\gamma_{kn}(\omega)}$$

where for abbreviation

$$d_k(\omega) \equiv \prod_{t=1}^{T_k} \left(\frac{C_{kt}}{\omega_{kt}} \right)^{\omega_{kt}} \qquad k = 0, 1, \ldots, K$$

and

$$\gamma_{kn}(\omega) \equiv \sum_{t=1}^{T_k} \alpha_{ktn} \omega_{kt} \qquad n = 1, \ldots, N$$

There is strict equality between the left and right members if and only if

$$C_{kt} \prod_{n=1}^{N} x_n^{\alpha_{ktn}} / \omega_{kt} = p_k(\mathbf{x}) \qquad k = 0, 1, \ldots, K$$

Rearrangement of these equations gives the reason the ω_{kt} are called dual *weights*.

$$\omega_{kt} = \frac{t_{kt}}{p_k} \qquad k = 0, 1, \ldots, K$$

That is, strict equality holds if and only if every dual weight ω_{kt} for the kth posynomial equals the fraction of p_k accounted for by the tth term t_{kt}.

Since no constraint function p_k can exceed unity, each can be raised to any nonnegative power without ever being greater than unity. That is, for every nonnegative number $\lambda_k \geq 0$, it is true that $p_k^{\lambda_k} \leq 1$ for $k = 1, \ldots, K$. Therefore

$$1 \geq p_k^{\lambda_k} \geq \left[d_k(\omega) \prod_{n=1}^{N} x_n^{\gamma_{kn}(\omega)} \right]^{\lambda_k}$$

$$= \left[d_k(\omega) \right]^{\lambda_k} \prod_{n=1}^{N} x_n^{\lambda_k \gamma_{kn}(\omega)}$$

A lower-bounding function on the objective $p_0(\mathbf{x})$ is constructed as follows.

$$p_0(\mathbf{x}) \geq p_0 \prod_{k=1}^{K} p_k^{\lambda_k} \geq d_0(\omega) \prod_{k=1}^{K} \left[d_k(\omega) \right]^{\lambda_k} \prod_{n=1}^{N} x_n^{\gamma_n(\omega)}$$

where γ_n, the exponent of x_n, is

$$\gamma_n \equiv \gamma_{0n} + \sum_{k=1}^{K} \lambda_k \gamma_{kn}$$

These expressions are simplified by defining $\lambda_0 \equiv 1$ so that the sums and products can include the objective. Then let

$$d(\omega) \equiv \prod_{k=0}^{K} d_k^{\lambda_k} = \prod_{k=0}^{K} \prod_{t=1}^{T_k} \left(\frac{C_{kt}}{\omega_{kt}} \right)^{\lambda_k \omega_{kt}}$$

and

$$\gamma_n(\omega) = \sum_{k=0}^{K} \lambda_k \gamma_{kn} = \sum_{k=0}^{K} \sum_{t=1}^{T_k} \alpha_{ktn} \lambda_k \omega_{kt}$$

Then the lower-bounding function is the right member of

$$p_0(x) \geq d(\omega) \prod_{n=1}^{N} x_n^{\gamma_n(\omega)}$$

This lower-bounding function can be made independent of x by setting the exponents $\gamma_n(\omega)$ to zero.

$$\sum_{k=0}^{K} \sum_{t=1}^{T_k} \alpha_{ktn} \lambda_k \omega_{kt} = 0 \qquad n = 1, \ldots, N$$

These N bilinear equations, one for each design variable x_n, are called the *bilinear orthogonality conditions*. Let T be the total number of terms in the problem, including those in the constraints.

$$T \equiv \sum_{k=0}^{K} T_k$$

Then there are T dual weights ω_{kt} and K numbers λ_k that must satisfy the N orthogonality conditions as well as the $N+1$ normality conditions. Thus if $(T+K)-(N+K+1) = T-N-1$ dual variables are chosen numerically, say by matching them to the known primal weights for a given design, then all the remaining variables can be computed. In fact, if $T-N-1 = 0$, then all dual variables are completely determined without any arbitrary choices.

as in the first approximation to the cofferdam problem (Sec. 3.3). For this reason the quantity $(T-N-1)$, which is the number of degrees of freedom in the combined orthogonality and normality conditions, is called the number of "degrees of difficulty"—the more there are, the harder the problem is. Problems with zero degrees of difficulty have very elegant closed-form solutions. Moreover, approximations involving zero degrees of difficulty are often used to bound problems with several degrees of difficulty.

Since the multipliers λ_k only appear when multiplying dual weights, the lower bound and orthogonality conditions can also be expressed in terms of what are called *dual variables* δ_{kt}.

$$\delta_{kt} \equiv \lambda_k \omega_{kt} \qquad k=0,1,\ldots,K;\ t=1,\ldots,T_k$$

The N *orthogonality conditions* are linear in the dual variables.

$$\sum_{k=0}^{K} \sum_{t=1}^{T_k} \alpha_{ktn}\delta_{kt}=0 \qquad n=1,\ldots,N$$

Since λ_0 is defined to be unity, the normality condition for the objective is

$$\sum_{t=1}^{T_0} \omega_{0t} = \sum_{t=1}^{T_0} \delta_{0t} = 1$$

The other normality conditions become equations for computing the λ_k from the δ_{kt}.

$$\sum_{t=1}^{T_k} \delta_{kt} = \sum_{t=1}^{T_k} \lambda_k \omega_{kt} = \lambda_k \sum_{t=1}^{T_k} \omega_{kt} = \lambda_k \qquad k=1,\ldots,K$$

The lower-bound inequality is

$$p_0(\mathbf{x}) \geq \prod_{k=0}^{K} \prod_{t=1}^{T_k} \left(\frac{C_{kt}}{\omega_{kt}}\right)^{\delta_{kt}} = \prod_{k=0}^{K} \prod_{t=1}^{T_k} \left(\frac{C_{kt}\lambda_k}{\delta_{kt}}\right)^{\delta_{kt}}$$

The right member is called the *dual function* $d(\boldsymbol{\delta})$, and the geometric programming *duality relation* is

$$p_0(\mathbf{x}) \geq \min_{\mathbf{x}>0} p_0(\mathbf{x}) = \max_{\boldsymbol{\delta}>0} d(\boldsymbol{\delta}) \geq d(\boldsymbol{\delta})$$

where the vector δ of dual variables must also satisfy the orthogonality and normality conditions.

In general it may not be known in advance which primal constraints are active at the minimum. The preceding formulation can be extended to this case by adopting the convention that any constraint loose at the minimum have its multiplier λ_k set to zero. When this happens, all the dual variables δ_{kt} for this constraint vanish, although the dual weights ω_{kt} do not.

Since the orthogonality and normality conditions have the same form in both the constrained and unconstrained cases, the dual geometric programming method of Sec. 3.18 is immediately applicable to the constrained case, as long as all functions are posynomials. Dual geometric programming codes can, however, run into numerical problems when it is not known in advance which constraints are loose at the optimum, for loose constraints have all of their dual variables vanish simultaneously. It may be hard to decide numerically just when a constraint wants to loosen. The rest of the chapter deals with this problem, along the way demonstrating how to develop simple design procedures, lower bounds, and designs without resorting to a dual geometric programming code. This sets the stage for methods extendable to the more general case where negative terms appear and the constraints must be bounded below instead of above.

4.3. CONSTRAINT ACTIVITY

Let $p(\mathbf{x}) \leq 1$ be a primal constraint. At a given point $(\mathbf{x})_0$, either $p(\mathbf{x})_0 = 1$, in which case the constraint is said to be *tight* at $(\mathbf{x})_0$, or $p(\mathbf{x})_0 < 1$, in which case it is *loose* at $(\mathbf{x})_0$.

In constructing lower-bounding functions, one uses the fact that there exists, for every primal design $\mathbf{x}$, a nonnegative exponent $\lambda \geq 0$ such that $[p(\mathbf{x})]^\lambda = 1$. If the optimization analysis requires λ to be positive, then certainly the constraint must be tight. Formally, if $\lambda > 0$, then $p^\lambda = 1$ if and only if $p = 1$. On the other hand, if the optimization analysis requires λ to vanish, then whether the constraint is tight or loose is irrelevant, for in both cases $p^\lambda = 1$. Thus when $\lambda = 0$, the constraint is said to be *inactive*, since it has no influence on the optimization. Because a tight constraint could be inactive, any constraint for which $\lambda > 0$ is always called *active*.

The next example will show that when a tight constraint contains a variable appearing neither in the objective nor in any other active constraint, it is inactive and can be deleted from the problem. The example will demonstrate how monotonicity analysis can determine the activity of the constraints. When the activity of some constraints is not clear, both possibilities must be investigated. As in the example, activity cannot always be determined in advance, for it may depend on the values of the parameters. But when each case is easy to compute, systematic evaluation of a small number of cases may be better than a general procedure.

4.4. HYDRAULIC CYLINDER DESIGN

This example illustrates the construction of a simple *conditional* design procedure by monotonicity analysis. By "conditional" is meant that the option to perform certain calculations depends on the outcome of earlier ones. This approach is in this case easier than geometric programming, although the latter method is applicable.

A hydraulic cylinder, a device for lifting heavy loads as in a car hoist or elevator or for positioning light ones as in an artificial limb, is characterized by its inside diameter i, its wall thickness t, its fluid pressure p, and its wall stress s. Figure 4-1 shows a typical hydraulic cylinder.

Force and thickness are bounded below: $f \geq F$; $t \geq T$. Pressure and stress are bounded above: $p \leq P$; $s \leq S$. The inside diameter is not bounded except that, like all variables, it must be positive. Force, pressure, and inside diameter are related by

$$f = \frac{\pi i^2 p}{4}$$

For monotonicity analysis, let this relation be written more abstractly as

$$\varphi_1(i, f^{-1}, p) = 1$$

which says that the constraint function increases in i and p, while decreasing in f. Inside diameter, pressure, stress, and thickness are related by

$$\varphi_2(i^{-1}, p^{-1}, s, t) = 1$$

For example, the hoop stress formula is $s = ip/2t$.

Suppose the designer wishes a cylinder with a small outside diameter $i + 2t$. Then a lower-bounding function on this outside diameter (abbreviated o.d.) can be constructed with the help of the geometric inequality

$$\text{o.d.} \equiv i + 2t \geq K i^{\omega_1} t^{\omega_2} \qquad K \text{ is some positive constant}$$

for all $\omega_1 + \omega_2 = 1$ and $\omega_1, \omega_2 > 0$. Let μ_1 and μ_2 be real (positive, negative, or

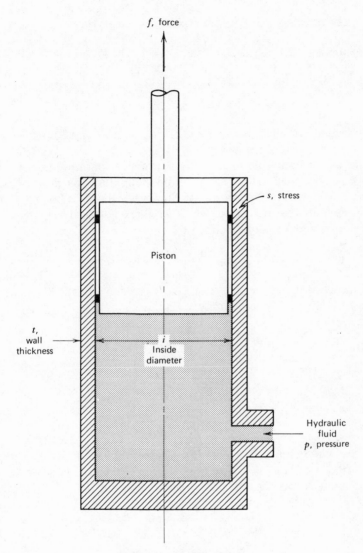

Figure 4-1 Hydraulic cylinder

111

zero) exponents such that $\varphi_1^{\mu_1} = 1$ and $\varphi_2^{\mu_2} = 1$. Then

$$\text{o.d.} \geq Ki^{\omega_1} t^{\omega_2} \varphi_1^{\mu_1} \varphi_2^{\mu_2}$$

$$= Ki^{\omega_1} t^{\omega_2} \left[\varphi_1(i, f^{-1}, p) \right]^{\mu_1} \left[\varphi_2(i^{-1}, p^{-1}, s, t) \right]^{\mu_2}$$

The immediate problem is to choose the signs of μ_1 and μ_2 that will keep the objective bounded below.

Begin by examining those variables that appear but once, namely f and s. Since f is bounded below, the lower-bounding function must be nondecreasing wrt f, which requires $\mu_1 \leq 0$. Similarly, since s is bounded above, the lbf cannot increase wrt s, which requires $\mu_2 \leq 0$. Thus nonnegative exponents λ_1 and λ_2 can be defined by

$$\lambda_1 \equiv -\mu_1 \geq 0; \lambda_2 \equiv -\mu_2 \geq 0$$

Then the equality constraints can be replaced by inequality constraint functions bounded above by unity.

$$\varphi_1^{-1} \equiv \varphi_1'(i^{-1}, f, p^{-1}) \leq 1$$

$$\varphi_2^{-1} \equiv \varphi_2'(i, p, s^{-1}, t^{-1}) \leq 1$$

The lbf may now be written

$$\text{o.d.} \geq Ki^{\omega_1} t^{\omega_2} \left[\varphi_1'(i^{-1}, f, p^{-1}) \right]^{\lambda_1} \left[\varphi_2'(i, p, s^{-1}, t^{-1}) \right]^{\lambda_2}$$

Since i is not bounded, it must cancel out of the lbf, which can only happen if $\lambda_1 > 0$, for ω_1 is known to be positive. Hence φ_1' must be active at the minimum. Moreover, the activity of φ_1' implies that the lbf increases in f, and so it must achieve its lower bound F.

$$f_\star = F$$

The designer now knows to design the cylinder to exert the least force allowed by the specifications. The force constraint is written

$$\varphi_1' \lessgtr 1$$

to symbolize that it is directed and active at the minimum.

It is not yet known whether φ_2' is active or not, so both cases must be examined. First, if the stress constraint φ_2' is inactive, then $\lambda_2 = 0$ and the lbf increases wrt p. Hence thickness and pressure achieve their bounds at the minimum o.d.

$$t_\star = T \text{ and } p_\star = P \qquad if \qquad \lambda_2 = 0$$

Moreover, the stress s vanishes from the lbf, so its upper bound must either be loose ($s < S$) or inactive. Only four variables are left. Three of these, f, p, and t, are known, and there is one active constraint that determines the fourth i. This happens to be solvable, so

$$i_\star = 2\left(\frac{F}{\pi P}\right)^{1/2} \quad if \quad \lambda_2 = 0$$

The resulting unique solution must be substituted into φ_2' to find $s_\star$. If $s_\star \leq S$, then the solution is optimal as well as feasible, and the case $\lambda_2 > 0$ need not be investigated. But if $s_\star > S$, then the case $\lambda_2 = 0$ has been proven infeasible, and only the possibility $\lambda_2 > 0$ remains.

When $\lambda_2 > 0$ the stress constraint is assumed active. This implies $s = S$. To be specific, let the hoop stress formula be used as an example; more complicated relations with the same monotonicity may be needed in a real situation. The problem is then to minimize $c = i + 2t$ subject to $(4F/\pi)i^{-2}p^{-1} \leq 1$ and $(2S)^{-1}ipt^{-1} \leq 1$. The other bounding constraints, $p \leq P$ and $t \geq T$, will be temporarily ignored, since the resulting problem will have zero degrees of difficulty. The solution is therefore easy to obtain and readily checked for feasibility wrt the bounds.

The orthogonality conditions are

$$
\begin{array}{llll}
i: & \omega_{01} & -2\lambda_1 + \lambda_2 = 0 \\
p: & & -\lambda_1 + \lambda_2 = 0 \\
t: & \omega_{02} & -\lambda_2 = 0
\end{array}
$$

These, together with the normality condition

$$\omega_{01} + \omega_{02} = 1$$

have a unique solution $\omega_{01} = \omega_{02} = \lambda_1 = \lambda_2 = \frac{1}{2}$. Hence

$$c \geq \left(\frac{1}{\frac{1}{2}}\right)^{1/2}\left(\frac{2}{\frac{1}{2}}\right)^{1/2}\left(\frac{4F}{\pi}\right)^{1/2}\left(\frac{1}{2S}\right)^{1/2} = 4\left(\frac{F}{\pi S}\right)^{1/2} \equiv \tilde{c}$$

The tilde ($\sim$) indicates that this cost may not correspond to a feasible design, although if it does, it is the minimum.

The corresponding value of i is found from the first term of the objective

$$\tilde{i} = \omega_{01}\tilde{c} = \frac{1}{2}\left(4\left(\frac{F}{\pi S}\right)^{1/2}\right) = 2\left(\frac{F}{\pi S}\right)^{1/2}$$

The second term gives

$$2\tilde{t} = \tilde{i}$$

from which

$$\tilde{t} = \left(\frac{F}{\pi S}\right)^{1/2}$$

Substitution of these into either constraint gives

$$\tilde{p} = S$$

If $\tilde{t} \geq T$ and $\tilde{p} \leq P$, then $\tilde{x} = x_\star$, the global minimum of the origina¹ problem. The conditions for feasibility of $\tilde{x}$ can be expressed entirely as functions of the problem parameters.

$$\tilde{t} = \left(\frac{F}{\pi S}\right)^{1/2} \geq T$$

or

$$T\left(\frac{\pi S}{F}\right)^{1/2} \leq 1$$
$$\tilde{p} = S \leq P$$

or

$$SP^{-1} \leq 1$$

The situations where the pressure and thickness constraints are violated must also be examined. First suppose $\tilde{t} < T$. Since c is minimum (although infeasible) at $\tilde{t}$, it strictly increases wrt t for all $t > \tilde{t}$, including the feasible region $t \geq T$. Therefore c is minimum in the feasible region when $t = T$. With t fixed, there are no degrees of freedom left, and so i and p are just the solutions of the constraints.

$$i = \frac{2F}{\pi ST} \qquad p = \frac{\pi S^2 T^2}{F}$$

If this pressure is feasible, that is, if $P^{-1}(\pi S^2 T^2 / F) \leq 1$, then this solution minimizes c. If, on the other hand, $p > P$, then since p increases with t (to see this, replace T by t), $p_\star = P$, which makes $i_\star = 2(F/P)^{1/2}$ and $t_\star = (FP/S^2)^{1/2}$ to satisfy the constraints.

Next suppose $\tilde{\imath}$ is feasible but not $\tilde{p}$. By similar reasoning it follows that the minimizing value of p must be P, the upper bound. The force constraint immediately gives $i = 2(F/\pi P)^{1/2}$. Substitution of this into the stress constraint yields $t = (FP/\pi S^2)^{1/2}$. If this thickness is feasible, i.e., if $(FP/\pi S^2)^{1/2} \geq T$ or $T(\pi S^2/FP)^{1/2} \leq 1$, then it is optimal. Its infeasibility would indicate that no feasible solution existed.

4.5. CONDITIONAL DESIGN PROCEDURE

Although the individual cases analyzed in the hydraulic cylinder example were simple, the overall analysis may have seemed complicated because of the numerous cases. Yet a particular design may not require examining all, or even most, cases, for with luck the first might prove optimal. Figure 4-2 is a flow diagram for a design procedure based on the preceding analysis. It is called a *conditional* procedure because the order of computing depends on the results of earlier tests. Notice that each test is quite simple, easily performed on a pocket calculator or for that matter on a slide rule. If many hydraulic cylinders are to be designed for a range of services, i.e., parameters, the designer may wish to program the entire procedure, either for a large computer or even for a programmable

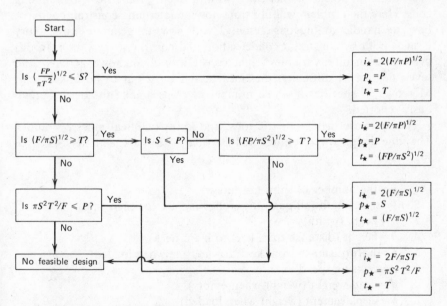

Figure 4-2 Conditional design for minimum diameter hydraulic cylinder

calculator. Many designers would rather not program the logic, preferring to use the chart to guide the investigation.

Conditional design procedures are therefore a natural extension of the well-known design formula concept. Even in this problem, which has the proper form for geometric programming, the conditional procedure is better, because it circumvents certain numerical difficulties suffered by geometric programming when it is unclear in advance which constraints are active.

4.6. MERCHANT FLEET DESIGN

In the hydraulic cylinder problem, monotonicity analysis proved there to be but two patterns of constraint activity at the optimum. When there are more constraints, the number of possibilities may at first appear impractically large. Yet an extension of monotonicity analysis can reduce this number to manageable proportions, as the next example illustrates. The example also shows how to generate lower bounds and feasible designs intelligently in a realistically large design problem.

The problem is to design a fleet of ships to transport a given amount of cargo within a certain time. Folkers formulated and solved this entirely posynomial problem numerically with a dual geometric programming code. Here the problem will illustrate how to determine constraint activity, how to handle ambiguous activity, and how to generate satisfactory designs both by constraint condensation and constraint relaxation. In this problem with nine variables, eight constraints of unknown activity, and seven degrees of difficulty, the third design turns out to be satisfactory. Moreover, it specifies a whole number of ships, unlike the design previously reported.

The units are meters, metric tons, and knots (nautical miles per hour). The nine design variables, denoted as usual by lower case letters, are (see Fig. 4-3):

1. n = the number of ships (an integer)
2. l = ship length (all quantities following are for each ship)
3. b = beam (width)
4. h = height (distance from keel to main deck)
5. t = draft (distance from keel to waterline when loaded)
6. v = cruising speed
7. d = dead weight (weight when empty)
8. δ = displacement (weight when loaded)
9. u = service factor (fraction of time at sea)

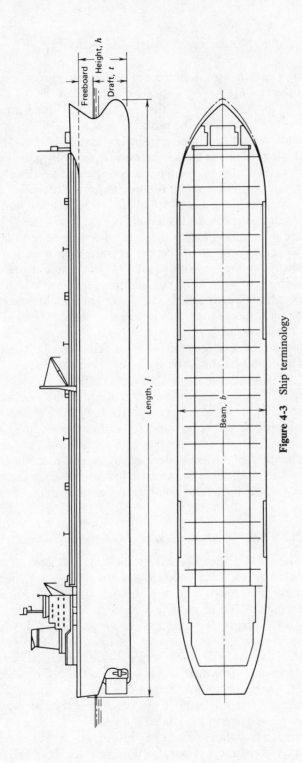

Figure 4-3 Ship terminology

The objective function to be minimized is the lifetime cost p_0, the sum of the hull cost $t_{01} \equiv 160n(lbh)^{0.9}$, the engine cost $t_{02} \equiv 2n\delta^{0.48}v^{2.16}$, and the fuel cost $t_{03} \equiv 0.8\delta^{0.48}v^{2.16}un$. There are three equality and five inequality constraints, and all variables must be positive.

The three normalized equalities are conservation equations. The first, p_1, apportions the total time for the job between $t_{11} = \frac{2}{3}(10^{-6})duv$, the fraction of time spent in port, and $t_{12} \equiv u$, the fraction spent at sea. The second, p_2, based on Archimede's Principle for floating objects, distributes the mass of the water displaced among $t_{21} \equiv 0.4(lbh)^{0.9}\delta^{-1}$, the fraction due to cargo, $t_{22} \equiv d\delta^{-1}$, the fraction due to dead weight, and $t_{23} \equiv 0.02\delta^{-0.52}v^{2.16}$, the fraction due to the engines. The third, p_3, is a normalized form of a marine architecture relation between the approximate shape of a hull and the cruising speed. It is the sum of $t_{31} \equiv 0.93\delta(lbt)^{-1}$ and $t_{32} \equiv 0.93l^{-1/2}v$. These equations are written in the unique form for which they are normalized equalities, but at this point it is not known whether they can be replaced by inequalities. Neither has it been established if any of them are really active at the optimum.

The five inequalities are posynomials bounded above by unity. One, p_4, acts as a lower bound on the amount of cargo to be delivered; it is a sum of $t_{41} \equiv 2.75(10^6)(nduv)^{-1}$ for the cargo and $t_{42} \equiv 2.1(10^{-3})d^{-1}\delta^{2/3}v^2$ for the fuel. Another, p_5, bounds the freeboard, the height of the main deck above the water, for safety; it is the sum of $t_{51} \equiv 1.9(10^{-3})l^{1.43}h^{-1}$ and $t_{52} \equiv th^{-1}$. A third bounds the draft/beam ratio for stability: $p_6 \equiv 2tb^{-1}$. A fourth bounds the length/height ratio for strength: $p_7 \equiv 0.07lh^{-1}$. And the last bounds the fraction of the ship devoted to cargo: $p_8 \equiv 3d(lbh)^{-1}$. The activity, or lack of it, of these inequalities has not yet been established. A monotonicity analysis is therefore the first order of business.

4.7. MONOTONICITY ANALYSIS

To determine the activities of the constraints, as well as the directions of the inequalities that could correctly replace the equations, a monotonicity analysis will be performed. The size of this problem requires a more compact format than the previous examples needed. For this it is convenient to use the matrix of exponents shown in Table 4-1, which has a column for each term t_{mt} and a row for each variable. Recorded are the exponents of the variables in these terms. Since these exponents are the coefficients of nonnegative dual variables δ_{mt} in the orthogonality conditions, one can imagine each row as representing a linear function in the dual variables that must vanish at the optimum. This will show which variables must be nonzero in a bounded problem.

The analysis is sequential in that early results generate later ones. At all times one must keep track of which constraint activities and directions

Table 4-1   Fleet Design Exponents

Vari-
able   Term Index

→	01	02	03	11	12	21	22	23	31	32	41	42	51	52	6	7	8
n	1	1	1								−1						
l	0.9					0.9			−1	−0.5			1.43			1	−1
b	0.9					0.9			−1						−1		−1
h	0.9					0.9							−1	−1		−1	−1
t									−1					1	1		
v		2.16	2.16	1				2.16		1	−1	2					
d				1			1				−1	−1					1
δ		0.48	0.48			−1	−1	−0.52	1			0.67					
u			1	1	1						−1						

have been established. In the beginning, only the directions of the constraints p_4 through p_8 are known; equivalently, dual variables δ_{41} through δ_{81} are known to be nonnegative, as are the dual variables δ_{01}, δ_{02}, and δ_{03} for the objective terms.

Consider first the top line, corresponding to the orthogonality condition for n, the number of ships. This can be written $\delta_{01} + \delta_{02} + \delta_{03} - \delta_{41} = 0$. Since $\delta_{01} + \delta_{02} + \delta_{03} > 0$, it follows that $\delta_{41} > 0$, which means that the cargo constraint must be active at the optimum to bound the number of ships ($p_4 \leqq 1$ and $\delta_{42} > 0$). This could have been deduced without writing the equation merely by noting that δ_{41} has the only negative sign in row n, and there are positive entries in the objective columns. The reader can verify that the activity of p_4 also bounds v, d, and u.

Line t proves that the shape-speed equation can be replaced by an inequality bounding it from above ($p_3 \leqq 1$), since both δ_{52} and δ_{61} are known to be nonnegative. Then line δ shows that the buoyancy constraint can be replaced by an active upper bounding inequality ($p_2 \leqq 1$), for only by having δ_{21}, δ_{22}, and δ_{23} positive can the displacement be bounded when $\delta_{31} \geq 0$ and $\delta_{42} > 0$.

More complicated logical relations involving several constraints can also be deduced. For example, line h implies that at least one of the constraints p_5, p_7, or p_8 must be active to bound the height. Before seeking such complicated relationships, however, let us look for simpler ones arising when terms are neglected.

The terms small enough to risk neglecting in a first approximation are the four involving the engines and the fuel, since these would account for only a small fraction of the displacement in a good design. Imagine then that terms t_{02}, t_{03}, t_{23}, and t_{42} are all neglected. For the simplified problem to be bounded, line δ shows that the form-speed relation must be active ($p_3 \leqq 1$), whereas line u proves that the time constraint can be replaced by an active upper bound ($p_1 \leqq 1$). Thus neglecting small terms may be viewed as a strategem not only for reducing degrees of difficulty and simplifying computations, but also as a way to identify constraints likely to be active at the optimum.

The first four constraints are now directed and known to be active, at least for the simplified problem. Moreover, lines h and t show that either p_5 is active, or else p_6 together with either p_7 or p_8 must be active. Any of these combinations would have 10 terms in nine variables, for zero degrees of difficulty. One of these will be solved to generate a base case that can be tested for feasibility with respect to the constraints temporarily assumed inactive. Since the freeboard constraint p_5 bounds both draft and height, it will be *assumed* active to generate a bounded zero degree of difficulty problem.

4.8. ORTHOGONALITY ANALYSIS

By neglecting the engine and fuel terms and assuming the freeboard constraint active, the following zero degree of difficulty posynomial geometric program is obtained: minimize $p_0' \equiv t_{11}$ subject to the active inequalities $p_1 \leqq 1, p_2' \equiv t_{21} + t_{22} \leq 1, p_3 \leqq 1, p_4' \equiv t_{41} \leqq 1, p_5 \leqq 1$, and the assumed inactive inequalities $p_6 < 1$, $p_7 < 1$, and $p_8 < 1$. Since the last three inequalities are said to be *relaxed*, this problem is called a *relaxation* of the original problem in eight constraints. Remember that in this relaxed problem, all of the dual variables $\delta_{02}, \delta_{03}, \delta_{23}, \delta_{42}$ have been set to zero, as have δ_{61}, δ_{71}, and δ_{81} for the relaxed constraints. Despite all the analysis that has been performed, there is still no guarantee that the condensed and relaxed problem is bounded. For this reason the other dual variables must be found, not as constants, but as functions of those duals temporarily set to zero.

Solving the linear orthogonality conditions together with the normality condition $\delta_{01} + \delta_{02} + \delta_{03} = 1$ gives the following 10 functions:

$$\delta_{01} = 1 \quad\quad - \delta_{02} \quad - \delta_{03}$$
$$\delta_{11} = 1 - 2.16\delta_{02} - 1.16\delta_{03} - 2.16\delta_{23}, \quad - 2\delta_{42} - 7.72\delta_{61} + 0.11\delta_{71}$$
$$\delta_{12} = \quad 2.16\delta_{02} + 2.16\delta_{03} + 2.16\delta_{23} \quad + 2\delta_{42} + 7.72\delta_{61} - 0.11\delta_{71}$$
$$\delta_{21} = 9 - 25.8\delta_{02} - 15.8\delta_{03} - 26.8\delta_{23} - 23.1\delta_{42} - 87.2\delta_{61} + 0.14\delta_{71}$$
$$\delta_{22} = \quad 2.16\delta_{02} + 1.16\delta_{03} + 2.16\delta_{23} \quad - \delta_{42} + 7.72\delta_{61} - 0.11\delta_{71} - \delta_{81}$$
$$\delta_{31} = 9 - 24.1\delta_{02} - 15.1\delta_{03} - 24.1\delta_{23} - 21.0\delta_{42} - 79.5\delta_{61} \quad\quad - \delta_{81}$$
$$\delta_{32} = \quad\quad\quad\quad\quad\quad \delta_{03} \quad\quad\quad\quad\quad\quad + 7.72\delta_{61} - 0.11\delta_{71}$$
$$\delta_{41} = 1$$
$$\delta_{51} = \quad\quad\quad\quad\quad\quad\quad\quad\quad\quad\quad\quad 2\delta_{61} - 0.75\delta_{71}$$
$$\delta_{52} = 9 - 24.1\delta_{02} - 15.1\delta_{03} - 24.1\delta_{23} - 21.0\delta_{42} - 80.5\delta_{61} - 0.06\delta_{71} - \delta_{81}$$

They show immediately that the dual variables on the right cannot all be zero as assumed, for if they were, then δ_{12}, δ_{22}, δ_{32}, and δ_{51} would all vanish. This would correspond to an absurd design with infinitely long but weightless ships that never leave port. However, a reasonable design can be obtained, even with the engine and fuel terms δ_{02}, δ_{03}, δ_{23}, and δ_{42} neglected, by making δ_{61} positive. This works because δ_{61} has a positive coefficient in every one of the functions δ_{12}, δ_{22}, δ_{32}, and δ_{51}, which is not true for δ_{71} and δ_{81}. Thus the stability constraint must be active ($p_6 \leqq 1$) to bound the problem, and the expression for δ_{51} shows that this in turn requires activity of the freeboard constraint ($p_5 \leqq 1$) as originally assumed. The exponents of the problem have now established the activity of the first six constraints, provided that the engine and fuel terms are really negligible.

In general the procedure, called *orthogonality analysis*, is to solve the orthogonality and normality conditions for a number of dual variables equal to the number of design variables plus one. The corresponding terms are said to be *dominant* because they are expected not to be negligible. Examination of the resulting functions of the dual variables remaining can, as in the example, reveal if other constraints must be active or other terms made dominant to preserve boundedness. In simple problems orthogonality analysis reduces to monotonicity analysis; in larger ones it can yield valuable insights and simplifications depending only on the exponents but not on the parameters of the model. For example, in the fleet design problem it can now be asserted that when engine and fuel effects are neglected, the optimal design will be restricted critically by time, buoyancy, the form/speed relation, the cargo requirement, freeboard, and stability. It is not known yet whether the remaining two constraints on strength and cargo space will be active; this may depend on the problem parameters. No more than three of the nine variables can be specified independently.

The smallest approximating problem that is still bounded has now been proven to have one degree of difficulty. The dual variable δ_{61}, upon which the others depend, must stay within a restricted range to prevent driving the other duals negative. The reader can verify that the most critical of these is δ_{21}, which becomes negative whenever $\delta > 9/87.2 = 0.103$. The next section shows how to make a reasonable choice of δ_{61} for generating a lower bound on the cost, as well as a base case design.

4.9. GENERATING A LOWER BOUND

Any positive choice of δ_{61} less than 0.103 will determine positive values of the dominant dual variables from which a lower bound can be computed. Since this will then be used to construct a design, the choice of δ_{61}

should reflect the realities of the situation. Although there is no physical basis for estimating δ_{61} directly, any proportion between dual variables, which would correspond to a ratio of the terms, would fix δ_{61}. A naval architect could make several intelligent guesses, but perhaps the easiest relation for a landlubber to understand is the weight distribution p_2. An unloaded ship looks like at least half of its hull is under water, meaning that the dead weight exceeds the maximum cargo load. Thus $t_{22} > t_{21}$, or $\delta_{22} = 7.72\delta_{61} > \delta_{21} = 9 - 87.2\delta_{61}$, from which $\delta_{61} > 0.095$. This brackets δ_{61} within 8%.

An even more precise estimate uses the fact that the number of ships must be a whole number. Since $\delta_{41} = 1$, it follows that $duv = 2.75(10^6)/n$. Hence $\delta_{11} = \frac{2}{3}(10^{-6})duv = 1.83/n = 1 - 7.72\delta_{61}$. Thus $0.095 < \delta_{61} < 0.103$ implies $6.8 < n < 8.9$. There are in fact only two integer possibilities for n—either seven or eight ships. Let eight be chosen, it being slightly nearer the middle of the interval. Notice that this number could be viewed as determined entirely by the exponents together with the cargo requirement and the coefficient for the fraction of time needed to load and unload in port.

When $n = 8$, $\delta_{61} = 0.100$, which determines the other dominant dual variables as $\delta_{11} = 0.229$, $\delta_{12} = 0.771$, $\delta_{21} = 0.293$, $\delta_{22} = 0.771$, $\delta_{31} = 1.064$, $\delta_{32} = 0.771$, $\delta_{41} = 1$, $\delta_{51} = 0.200$, $\delta_{52} = 0.964$. A lower bound on the cost is provided by computing the dual function as shown.

$$c \geq 160\left[2.75(10^6)\right]\left[\frac{667(10^3)}{0.229}\right]^{0.229}(0.771)^{0.771}\left(\frac{0.4}{0.293}\right)^{0.293}(0.771)^{0.771}$$

$$\times (1.064)^{1.064}\left(\frac{0.93}{1.064}\right)^{1.064}\left(\frac{0.93}{0.771}\right)^{0.771}(1.835)^{1.835}$$

$$\times \left[\frac{1.9(10^{-3})}{0.200}\right]^{0.200}(0.964)^{-0.964}(1.163)^{1.163}2^{0.100}$$

$$= 66.1(10^6)$$

The third figure is not significant, since many of the coefficients are given only to one or two figures. It is retained merely to reduce rounding error in comparing subsequent designs. The satisfaction threshold will be taken in this problem to be 1%, which, although corresponding to an uncertainty of $700,000, is severe considering the inaccuracy of the coefficients. The active constraints will be considered satisfied if the constraint functions are 1.00, that is, unity to two decimal places.

4.10. BASE CASE

The next task is to generate a base case design. When there are no degrees of difficulty, as in previous examples, the invariance conditions can be used for this, but now there is a degree of difficulty. Thus the invariance conditions will not be consistent except at the exact optimum. This situation is circumvented by condensing one of the two-term constraints into a single term, for this gives exactly as many constraint terms as variables and makes the set consistent. The objective function is not used to generate the design, its value being computed only as an accuracy check. Because the condensed function is a lower bound on the original constraint function, the design will be slightly infeasible unless the weights chosen happen to be exactly optimal. Consequently, the objective function for the design may be less than the lower bound which generated it. If the infeasibility is detectable within the second decimal place, the constraint must be recondensed and a new design computed. Otherwise the first design can be accepted if the cost matches the lower bound to within the accuracy threshold.

Let the buoyancy constraint p_2 be chosen for condensation, using the weights $0.293/1.064 = 0.275$ and $0.771/1.064 = 0.725$ suggested by the dual solution. Then

$$1 \geq p_2 \geq \left[\frac{0.4(lbh)^{0.9}}{0.275} \right]^{0.275} \left(\frac{d}{0.725} \right)^{0.725} \delta^{-1}$$

$$= 1.400(lbh)^{0.248} d^{0.275} \delta^{-1}$$

The right member is now set to unity to obtain a design relation. The other eight design equations follow, identified by the terms that generated them.

$$t_{11} = 0.229 = \tfrac{2}{3}(10^{-6}) duv \qquad t_{12} = 0.771 = u \qquad t_{31} = \frac{1.064}{1.835} = 0.93\delta\,(lbt)^{-1}$$

$$t_{32} = \frac{0.771}{1.835} = 0.93 l^{-1/2} v \qquad t_{41} = 1 = 2.75(10^6)(nduv)^{-1}$$

$$t_{51} = \frac{0.200}{1.163} = 1.9(10^{-3}) l^{1.43} h^{-1}$$

$$t_{52} = \frac{0.964}{1.163} = th^{-1} \qquad t_{61} = 1 = 2tb^{-1}$$

The design, obtained easily by direct algebraic manipulation or by taking logarithms, solving the linear equations resulting, and then exponentiating, is given in Table 4-2. In all cases the final figure is not significant, being retained merely to reduce rounding error. The objective is $66.6(10^6)$,

Table 4-2   First Design

No. of Ships (n)	Length (l)	Beam (b)	Height (h)	Draft (t)	Speed (v)	Dead Weight (d)	Displace- ment (δ)	Service Factor (u)	Cost
8	206 m	37.4 m	22.6 m	18.7 m	6.48 knots	$68.8 \cdot 10^3$ tons	$89.6 \cdot 10^3$ tons	77.1 %	\$66.6 $\cdot$ 10^6

slightly higher than the lower bound of $66.1(10^6)$, but within the accuracy threshold.

The neglected engine and fuel terms are $t_{02} = 0.30(10^6)$, $t_{03} = 0.09(10^6)$, $t_{23} = 0.003$, and $t_{42} = 0.004$, which raises the cost to $67.0(10^6)$, just outside the satisfaction threshold. The constraints must also be checked for feasibility: $p_1 = 1$ by construction; $p_2 = 0.232 + 0.768 + 0.003 = 1.003$; $p_4 = 1.004$; $p_3 = p_5 = p_6 = 1$ by construction. This shows the condensed constraint to be satisfied, even though the weights have shifted noticeably.

Finally the relaxed constraints must be checked: $p_7 = 1.00$; $p_8 = 3d/lbh = 1.19 > 1$. The strength constraint p_7 is just active to two decimal places, but the cargo space constraint p_8 is violated. This first design must therefore be discarded as infeasible.

4.11.   A FEASIBLE DESIGN

The cargo space constraint must now be added to the active set, adding another degree of difficulty to the problem. The barely active strength constraint will remain relaxed until the effect of activating p_8 has been determined. To compensate for the new degree of difficulty, another two-term constraint must be condensed: the freeboard constraint p_5 will be chosen because its weights did not change much between the dual solution and the first design. Thus the two design relations derived from t_{51} and t_{52} are replaced by the single equation

$$\left(\frac{1.9(10^{-3})l^{1.43}}{0.172} \right)^{0.172} \left(\frac{t}{0.828} \right)^{0.828} h^{-1} = 0.539 l^{0.246} t^{0.828} h^{-1} = 1$$

obtained by condensing p_5. Since the weights for buoyancy constraint p_2 did shift, it is recondensed with the weights 0.232 and 0.768 of the first design. The new equation is

$$1.390(lbh)^{0.209} d^{0.768} \delta^{-1} = 1$$

replacing the earlier one in these variables. These two equations, together

Table 4-3   First Feasible Design

n	l	b	h	t	v	d	δ	u	Cost
8	274	30.8	21.2	15.4	7.50	59.5	81.0	77.1	68.2

Table 4-4   Terms for First Feasible Design

Term	Posynomial (m)								
t	0	1	2	3	4	5	6	7	8
1	$68.2 \cdot 10^6$	0.229	0.263	0.580	1	0.274	1	0.90	1
2	0.28	0.771	0.735	0.422	0.003	0.726	—	—	—
3	0.09	—	0.004	—	—	—	—	—	—
Total	$68.6 \cdot 10^6$	1.000	1.002	1.002	1.003	1.000	1	0.90	1

with the seven derived earlier from the remaining terms, give the design in Table 4-3. The new design is feasible, as shown in Table 4-4, but its cost of $68.6(10^6)$ is 2.5% above the lower bound. Another iteration is needed.

4.12. TERMINATION

It is not yet known whether the 2.5% gap between lower bound and feasible design is due to an inferior design, an inferior lower bound, or both. One would expect to improve the lower bound by letting the dual variable for the newly active cargo space constraint become positive. Allowing the number of ships to be noninteger would also help.

The new condensed and relaxed problem has two degrees of difficulty, but at least there is a feasible design to give some idea what the weights should be. For example, the buoyancy constraint p_2 has weights 0.264 and 0.736 for its first two terms, the dominant ones. Dividing the first by the second gives a ratio of the dual variables δ_{21}/δ_{22}, each a linear function of the unknowns δ_{61} and δ_{81}. Expansion of this ratio gives

$$0.736\delta_{21} = 0.736(9 - 87.2\delta_{61}) = 0.264\delta_{22} = 0.264(7.72\delta_{61} - \delta_{81}).$$

Rearrangement yields the linear equation

$$66.2\delta_{61} - 0.264\delta_{81} = 6.62$$

Similar use of form/speed constraint p_3 and freeboard constraint p_5 gives, respectively, the following two additional equations.

$$37.9\delta_{61} + 0.421\delta_{81} = 3.79$$
$$23.5\delta_{61} + 0.274\delta_{81} = 2.47$$

These three equations in two unknowns are of course inconsistent, since the design from which they are derived is not optimal. Still one can find the solution that fits them in the least-squares sense of minimizing the sum of squared errors between the right and left members. This will give an orthogonal set of dual variables generating weights near those of the last design, which is known to be nearly optimal. The least squares solution can be proven to be obtained by the following procedure. Let the set of linear equations be written in matrix form as $M\delta = b$, where M is the coefficient matrix (3×2 in the example), δ the vector of dual variables to be determined (2×1 in the example), and b the vector of constants (3×1 in the example). The least-squares solution is then the solution to the square system obtained by multiplying both sides on the left by M^T.

$$(M^T M)\delta = M^T b$$

In the example, this system yields two scalar equations in the two unknowns:

$$6371\delta_{61} + 4.92\delta_{81} = 640.0$$

$$4.92\delta_{61} + 0.32\delta_{81} = 0.525$$

The matrix multiplications are easily performed using the statistical functions on a calculator. The least-squares solutions are $\delta_{61} = 0.100$ and $\delta_{81} = 0.097$; notice that the first did not change. These generate the dual variables used in the following computation of a new lower bound.

$$c \geq 160\left(\frac{0.667(10^{-6})}{0.224}\right)^{0.224}\left(\frac{1}{0.776}\right)^{0.776}\left(\frac{0.4}{0.238}\right)^{0.238}\left(\frac{1}{0.679}\right)^{0.679}$$

$$\times (0.917)^{0.917}\left(\frac{0.93}{0.917}\right)^{0.917}\left(\frac{0.93}{0.776}\right)^{0.776}(1.693)^{1.693}\left(\frac{1.9(10^{-3})}{0.201}\right)^{0.201}$$

$$\times \left(\frac{1}{0.816}\right)^{0.816}(1.017)^{1.017}2^{0.100}3^{0.097}$$

$$= 66.8(10^6)$$

The new lower bound is 1% greater than the first one, narrowing the gap to only 1.5% between design and lower bound. Since this still exceeds the

Table 4-5   Final Design

n	l	b	h	t	v	d	δ	u	Cost
8	239 m	33.9 m	21.7 m	17.0 m	7.62 knots	$58.5 \cdot 10^3$ tons	$79.9 \cdot 10^3$ tons	77.1 %	$$67.2 \cdot 10^6$

Table 4-6   Terms for Final Design

Term	Posynomial m								
t	0	1	2	3	4	5	6	7	8
1	$67.2 \cdot 10^6$	0.229	0.263	0.540	1.000	0.220	1.003	0.770	1.000
2	0.29	0.771	0.732	0.459	0.004	0.783	—	—	—
3	0.01	—	0.005	—	—	—	—	—	—
Total	$67.5 \cdot 10^6$	1.000	1.000	0.999	1.004	1.003	1.003	0.770	1.000

satisfaction threshold, a new design will be generated from the new dual solution.

Table 4-5 gives the new design, obtained as before, and Table 4-6 shows it to be feasible within the significance of the coefficients of the problem. The total cost, allowing for quantities previously neglected, is $67.5(10^6)$, just 1% above the lower bound of $66.8(10^6)$. *Satis quod sufficit*, or, as the Cariocas say with thumbs on high, "*ótimo!*"

4.13.   SUPERSHIPS

The designer, whether of ships, shoes, or sealing wax, may find it interesting that a fleet of ships can be designed at all on a manual calculator, for then the approach holds promise for other reasonably large problems. The way in which a very small amount of physical "intuition," that is, experience of the world, was amplified by the mathematics into a very good starting design may inspire specialists to reexamine their own technologies for similar useful ideas. The twin requirements that the orthogonality conditions be satisfied, but that the problem remain bounded, seem to have ruled out all but a few possibilities. Fortunately, some of these, unlike the exact optimum, are easy to find.

Much of the value of this analysis for the naval architect goes beyond the original and important contribution of Folkers; namely, that the fleet-design problem can be formulated and solved as a posynomial geometric programming problem. One additional insight is how to round off the number of ships to get a starting design. The upper bound $n_\star \leq 8.9$

derived for the optimal number of ships, together with the successful rounding of this bound down to the nearest integer to generate a good design, permits quick elimination of uneconomical situations. This relation can be generalized parametrically by letting C_{11} be the coefficient of the port time term $t_{11}(0.667(10^{-6})$ in the example) and C_{41} be the coefficient of the cargo requirement term t_{41} [$2.75(10^6)$ in the example]. Then

$$n_\star = \frac{C_{11}C_{41}}{1 - 7.72\delta_{61}} \leq \frac{C_{11}C_{41}}{1 - 7.72(0.103)} = 4.88\,C_{11}C_{41}$$

The optimal service factor u can be bounded similarly, a good exercise. For the value of C_{11} in the example, this leads to the conclusion that each ship in the optimal fleet will carry at least $308(10^3)$ tons per unit time. Determining the dimensions of this smallest possible optimal ship is left as an instructive exercise for other armchair admirals (Exercise 4-15). Notice that since all the other constraint coefficients are constants, determined either by immutable physical laws or by the design rules of naval architecture, this smallest ship design will depend only on the port time coefficient C_{11}. Thus the cargo handling equipment, the type of cargo, and the port facility completely determine the fleet design, an observation of potential value in a systems study involving design of the port as well as the fleet.

The fact that not all constraints need be active—specifically, that the strength and cargo space constraints may be loose in a particular design —is perhaps as interesting as the obverse fact that time, buoyancy, form, cargo capacity, freeboard, and stability all shape the optimal design. Another intriguing point is the negligibility of the engine and fuel terms, at least as far as determining ship dimensions is concerned. These terms are of course still needed for cost estimation. Remarkably, the shipbuilding cost coefficient C_{01} (160 in the example) does not affect the design at all, as long as it is large enough to make the engine and fuel costs negligible by comparison. If it were different, the cost would change in proportion, but not the number or size of the ships. Thus the results are more universal than one might have at first suspected.

There is also an important mathematical consequence for the design theory developed in this book. The problem could not have been solved numerically by a geometric programming code, as did Kramers, if there had been any negative terms anywhere in the problem. However, the method of orthogonality analysis used here is not restricted to posynomials. Negative terms need only have the corresponding signs changed in the orthogonality table, since each line can be regarded as an optimality condition obtained by setting semilog derivatives to zero, rather than as an orthogonality condition for constructing a lower bound. The next two chapters will show in fact how to construct valid lower bounds when there are negative terms present.

The final implications of this analysis concern the general public, as represented hopefully by their maritime regulating agencies. The ships in this design are huge enough to qualify as superships (Noel Mostert cogently critiques the large petroleum tankship concept in his book *Supership*). To reduce the chance of ecological disaster induced by tanker accidents, Mostert has suggested several restrictions on tankship operations and design. One of these is that each ship have two engines to improve maneuverability and prevent total loss of power at sea. Despite the widespread occurrence of twin-screw vessels in other maritime services, some tanker operators have resisted this to take advantage of economies of scale in engine cost. But the present analysis has shown that, from an optimal design standpoint, the total engine cost is a negligible fraction of the cost of an optimal fleet. Ponder this paradoxical consequence of "*satis quod sufficit*": although two engines, each of half the power requirement, cost more than a single engine of full power, an optimal fleet of twin-screw ships is economically indistinguishable from an optimal fleet of less safe single-engine vessels. Any extra cost is swallowed up by uncertainty in the state of the art of naval architecture. In fact, adding a term to the cost to account for the risk of nautical accident would, if insurance and legal penalties reflect the true costs even roughly, certainly justify the second engine. Given the ecological damage of a tanker accident, governments could even afford to pay for the extra engine out of shipping taxes and port fees.

Mostert also worries about whether such long ships will break up in a storm. The analysis here gives comfort in showing slack in the strength constraint, meaning it could be made stricter with little or no economic effect. The strength relation $p_7 = 0.07lh^{-1} \leq 1$ is of course very approximate, and a more precise relation would provide more security. But the point is not so much that there exist good designs, numerically indistinguishable from the theoretical optimum, which are well inside the strength constraint. It is rather that the ships can be made stronger, and therefore safer, with little or no cost increase—by proper design.

A glance at the various designs shows but little change in cost for a large variation in length. This suggests that lengths short enough for good ship handling and mooring may not be uneconomical at all. Further study of this point may therefore be justified.

All these suggestions are, of course, only as good as the model from which they are derived. For the designer who is not a naval architect, they are intended merely to demonstrate the results of a typical design study using the brand of optimization and bounding theory peculiar to this book.

4.14. CONCLUDING SUMMARY

This chapter completes the development of geometric programming theory, which, in order to use the geometric inequality, must confine itself to posynomial objective and constraints. However, computer-oriented geometric programming algorithms were not used to design the hydraulic cylinder and the merchant fleet. Instead, a simple conditional design procedure was developed for the cylinder, permitting rapid design on a pocket calculator and giving engineering insight into how the optimum would change qualitatively as parameters change. Monotonicity analysis alone was enough to determine the basic pattern of constraint activity, and the subproblems generated, having either no degrees of freedom or no degrees of difficulty, had closed-form solutions.

The merchant fleet problem, with nine variables and eight constraints, yielded only partially to simple monotonicity analysis. Study of the orthogonality conditions themselves was needed. In this orthogonality analysis, a certain set of dominant dual variables, corresponding to large terms in the original primal problem, were expressed as linear functions of the remaining recessive duals, corresponding to negligible terms or supposedly inactive constraints. Reasonable guesses of which terms to neglect cut the starting problem down to zero degrees of difficulty, but the orthogonality analysis proved that another constraint had to be activated to keep the objective bounded. A lower bound for the resulting single-degree-of-difficulty problem was generated from dual variables corresponding to a whole number of ships with a reasonable proportion of cargo. Partial invariance gave a design violating one of the other constraints, but a feasible base case was generated by condensing a pair of invariance conditions to make room for the new constraint term. A least-squares fit to the orthogonality conditions gave a set of dual variables for a better lower bound, and a partial invariance produced a new design costing within 1% of the lower bound.

A designer in a hurry, with only one particular design problem to solve, would be better off getting a numerical answer from a dual geometric programming computer code. But if, as is usually the case, the designer is a specialist designing the same device under many different circumstances, the type of analysis used in this chapter is justified. In the long run it will save much computer time, keep the design under the designer's control, and most important, occasionally suggest fundamentally better designs.

NOTES AND REFERENCES

The references for Chapter 3 give extensions to the constrained case.

Folkers, J. S., "Ship Operation and Design" in Avriel, Rijckaert, and Wilde, pp. 221–226 (*op. cit.* Chapter 1).

Mostert, Noel, *Supership*, Knopf, New York, 1974.

EXERCISES

4-1.   Iterate the condensation at the end of Sec. 4.1.

4-2.   In the gravel sled problem of Sec. 3.16, express the volume variable $v \equiv lwh$ and reformulate the problem as one with four variables and a single active inequality constraint (Sec. 4.1).

*4-3.   Formulate your design project as minimizing a function subject to normalized inequality constraints. Indicate exactly where your model deviates from the posynomial form required by geometric programming. By changes of variable and rearrangement, see how close to the geometric programming form you can get (Sec. 4.2).

4-4.   Write Problem 4-2 in two forms. In the first, have v appear in the objective; in the second, write the objective only as a function of l, w, and h. Determine the appropriate exponent λ in each case and interpret the results in terms of the concept of constraint activity (Sec. 4.3).

4-5.   Suppose the horizontal perimeter of the gravel sled (Problem 4-2) must not exceed a parameter P. Develop a conditional design procedure (Secs. 4.4 and 4.5).

4-6.   Design a hydraulic cylinder: (a) for $F = 10^4$ kg, $P = 250$ kg/cm^2, $S = 5000$ kg/cm^2, $T = 3.0$ mm; (b) for a cheaper material having only half the allowable stress of the design in (a) (Sec. 4.5).

*4-7.   On your design project, if it is posynomial, perform a monotonicity analysis by constructing a table similar to that in Table 4-1, Sec. 4.7. Use the primitive form of your problem before any preliminary monotonicity analysis of the type of Chapter 2 has been performed. Verify your earlier results and see if any new constraint relations become apparent.

*4-8.   Redo Problem 4-7, neglecting the small terms noted in Problem 3-3 (Sec. 4.7).

*4-9.   If it is posynomial, perform an orthogonality analysis on your design project (Sec. 4.8).

*Project problem.

*4-10. If you had any success with Problem 4-9, generate a lower bound (Sec. 4.9).

*4-11. Generate a base case for your project by the methods of Sec. 4.10.

*4-12. If any constraints were violated in Problem 4-11, generate a feasible design as in Sec. 4.11.

4-13. Use the least-squares procedure of Sec. 4.12 on the cofferdam problem as stated in Sec. 3.9. Then compare the result with the Invariance Principle.

*4-14. If appropriate, try the least-squares procedure of Sec. 4.12 on your project.

4-15. Design the smallest possible optimal ship, as suggested in Sec. 4.13.

CHAPTER **5**

Profit

Income 21 shillings, expenses 20; result: happiness.
Income 19 shillings, expenses 20; result: misery.
a deliberate misquotation of Charles Dickens
(1840)

In *David Copperfield*, Dickens really had Mr. Micawber say, "Income ...20 shillings, expenses...19; result: happiness. Income...20 shillings, expenses...21; result: misery." This exhortation to be thrifty on a fixed income is certainly in the spirit of the preceding chapters, where the designer can achieve happiness only by minimizing cost. But of course there is another way—increase income, as in the misquotation. Realistically, the designer can often justify increased cost if the device, being more effective and therefore more valuable to consumers or clients, brings in more money. This is precisely the situation with energy-efficient home appliances, whose higher initial price is rapidly recovered through reduced utility costs.

Obviously one should maximize the difference between income and cost. But this simple subtraction brings mathematical problems because geometric programming can only deal with sums, not differences, of power function terms. The geometric inequality being valid only for positive variables, dual geometric programming cannot be used for profit maximization.

To the readers of this book, however, this is no great loss, since the preceding chapters never used dual geometric programming anyway. Condensation and partial invariance are still available, not only for generating improved designs, but even for using the geometric inequality on the cost,

133

although not on the income, to construct bounds. For maximization, these must be *upper* bounds of course. But with reasonable care, the principle of *satis quod sufficit* can be extended to profit maximization.

"Reasonable care" does, however, require more effort for profit maximization than it did for cost minimization. The methods of the earlier chapters could only converge to an optimum, a happy condition not guaranteed when there are negative terms. To understand why this happens, to recognize when things are going wrong, and to know how to get back on the right track requires more acquaintance with the fundamentals of optimization theory than before.

Fortunately, the geometric programming concepts of degrees of difficulty and invariance do extend to the optimization of *signomials*, the technical term for differences of posynomials. When there are many degrees of difficulty, the ease with which semilogarithmic second, as well as first, derivatives can be computed makes the well-known Newton-Raphson (N-R) method practical. This numerical direct search method is not used much at present not only because most engineers detest second derivatives, but also because the technique is often unstable. But the simplicity of second semilog derivatives of power functions refutes the former objection, and the constructive second-order optimization theory developed shows how to stabilize the temperamental N-R method. A posynomial being merely a special case of a signomial, this stabilized N-R method can also be applied to cost-minimization problems with too many degrees of difficulty for condensation and partial invariance.

5.1. PROFIT MAXIMIZATION OF A FERTILIZER PLANT

The next example deals with two design variables and no constraints. This time the objective, being profit, is to be maximized, and there are not only both negative and positive terms, but also several degrees of difficulty. The approach to this problem will be more rigorous than that used for the earlier examples, with precise definitions of such fundamental concepts as global and local minimum and maximum, as well as stationary point.

The design problem is to choose the pressure p and raw material feed rate m for a proposed fertilizer plant consisting of a compressor and a reactor. All figures following are on an annual basis, with the capital costs amortized over the expected lifetime of the plant. The gross income from selling the product is estimated as $360m^{0.80}$, the exponent reflecting the fact that demand increases slightly as the price decreases. The variable portion of the compressor capital cost is estimated at $6.27(10^{-6})m^{0.80}p^{2.30}$, while the variable reactor cost is $9.92(10^3)m^{0.90}p^{-0.90}$. In addition, the energy cost is $15.5(10^{-3})mp$, and the raw material cost is $0.650m$. There are also fixed costs for sales, advertising, overhead, labor, rent, and structures that total $1.37(10^6)$ money units. The net profit is therefore

$$n \equiv 360m^{0.80} - 6.27(10^{-6})m^{0.80}p^{2.30} - 9.92(10^3)m^{0.90}p^{-0.90}$$
$$- 15.5(10^{-3})mp - 0.650m - 1.37(10^6)$$

It is convenient to work only with the variable portion b of this net profit, so that $n \equiv b - 1.37(10^6)$ and

$$b \equiv 360m^{0.80} - 6.27(10^{-6})m^{0.80}p^{2.30} - 9.92(10^3)m^{0.90}p^{-0.90}$$
$$- 15.5(10^{-3})mp - 0.650m$$

The designer wishes to choose positive values of the design variables m and p making the net profit n, and consequently the variable profit b, as large as possible.

5.2. OPTIMIZATION

The fertilizer plant design problem can be expressed as what is called more generally an *optimization* problem. If $\mathbf{x}$ is a vector of design variables constrained to lie within some *feasible set* $\mathcal{F}$, and if $y(\mathbf{x})$ is some *objective function* depending on $\mathbf{x}$, then the *maximum value* of $y(\mathbf{x})$, written $y^\star$, is defined by $y(\mathbf{x}) \leq y^\star$ for all $\mathbf{x}$ in $\mathcal{F}$, with equality achieved at the *maximizer* $\mathbf{x}^\star$, which must be feasible

$$y(\mathbf{x}^\star) = y^\star$$

with

$$\mathbf{x}^\star \in \mathcal{F}$$

These relations are also written

$$\max_{\mathbf{x} \in \mathcal{F}} y(\mathbf{x}) = y(\mathbf{x}^\star) = y^\star$$

Here the maximizer $\mathbf{x}^\star$ appears implicitly; an explicit definition is

$$\mathbf{x}^\star = \arg \max y(\mathbf{x})$$

[read "the argument maximizing $y(\mathbf{x})$"].

Minimum value and *minimizer* are defined similarly, except that the inequality must be reversed and the star $\star$ made a subscript instead of a superscript. Thus $y_\star$ and $\mathbf{x}_\star$ are respectively the minimum value and minimizer.

$$y(\mathbf{x}) \geq \min_{\mathbf{x} \in \mathcal{F}} y(\mathbf{x}) = y(\mathbf{x}_\star) = y_\star$$

$$\mathbf{x}_\star = \arg \min y(\mathbf{x})$$

The words *optimum value* and *optimizer* refer to whichever case the designer (or "optimist") wishes to find. Thus if the objective is profit, the maximum is considered optimal, whereas if cost is the objective, the minimum is optimal. Incidentally, the *worst* cases are called the *pessimum value* and *pessimizer*. Although nobody deliberately engages in pessimization, there are published examples where, by mistake, the pessimum was found when an optimum was sought. The wise designer will be enough of an optimist to want the optimum, and enough of a pessimist to guard against errors.

When the feasible region $\mathcal{F}$ is open, the optimum may not exist. For example, let $y(\mathbf{x}) = x$, and let $\mathcal{F}$ be the set of all positive real numbers: $\mathcal{F} \equiv \{x : x > 0\}$. Then although $y(\mathbf{x}) > 0$ for all feasible $\mathbf{x}$, the point $x = 0$ is not itself feasible. Here $x = 0$ is the greatest lower bound on $y(\mathbf{x}) = x$. Another name for the greatest lower bound on $y(\mathbf{x})$ is the *infimum*. This is written

$$y(\mathbf{x}) \geq \inf_{\mathbf{x} \in \mathcal{F}} y(\mathbf{x})$$

In the example, however, equality cannot be achieved for feasible x. Thus an infimum always exists for a real-valued objective, although a minimum may not. Whenever the infimum occurs at a feasible point, it will be called a minimum rather than an infimum. The analogous concept for least upper bound is called the *supremum*.

$$y(\mathbf{x}) \leq \sup_{\mathbf{x} \in \mathcal{F}} y(\mathbf{x})$$

5.3. STATIONARY POINTS

Let $\mathbf{x}_0 \equiv (x_1, \ldots, x_N)_0^T$ be any feasible point ($\mathbf{x}_0 \in \mathcal{F}$) where all constraints are strict inequalities. Such a point is said to be *interior* or *locally unconstrained*. By this definition, there exists a feasible neighborhood of points $\mathbf{x}$

near $\mathbf{x}_0$ such that, if δ is a positive constant ($\delta > 0$), then

$$\left\{ \Sigma[x_n - (x_n)]_0^2 \right\}^{1/2} \equiv \left[\Sigma(\partial x_n)^2 \right]^{1/2} \equiv (\partial\mathbf{x}^T\partial\mathbf{x})^{1/2} < \delta.$$

Suppose $y(\mathbf{x})$ is differentiable in this neighborhood, and let the symbol $\partial y/\partial\mathbf{x}$ represent the row vector of first partial derivatives wrt each x_n.

$$\frac{\partial y}{\partial\mathbf{x}} \equiv \left(\frac{\partial y}{\partial x_1}, \ldots, \frac{\partial y}{\partial x_N} \right)$$

The numerical value of $\partial y/\partial\mathbf{x}$, evaluated at $\mathbf{x}_0$, is written

$$\left(\frac{\partial y}{\partial\mathbf{x}} \right)_0 \equiv \left(\left(\frac{\partial y}{\partial x_1} \right)_0, \ldots, \left(\frac{\partial y}{\partial x_N} \right)_0 \right)$$

In this notation, the first-order Taylor expansion of $y(\mathbf{x})$ in the neighborhood of $\mathbf{x}_0$ can be written $\partial y = (\partial y/\partial\mathbf{x})_0\partial\mathbf{x} + 0(\partial x_n^2)$ where $0(\partial x_n^2)$ represents a remainder of degree 2 or higher in the ∂x_n ($n = 1, \ldots, N$). A more compact notation is $\partial y \sim (\partial y/\partial\mathbf{x})_0\partial\mathbf{x}$ where the symbol $\sim$ indicates that equality occurs when the second order remainder $0(\partial x_n^2)$ is added.

The transpose of $(\partial y/\partial\mathbf{x})_0$ is a vector in $\mathbf{x}$-space, as are all vectors formed by multiplying $(\partial y/\partial\mathbf{x})_0^T$ by any nonzero real constant k. Consider the particular perturbations defined by $\partial\mathbf{x} = k(\partial y/\partial\mathbf{x})_0^T$. The change in y for this perturbation is $\partial y \sim (\partial y/\partial\mathbf{x})_0(\partial y/\partial\mathbf{x})_0^T k$. But the inner product of any vector with itself is a non-negative constant: $(\partial y/\partial\mathbf{x})_0(\partial y/\partial\mathbf{x})_0^T \equiv K \geq 0$. Therefore $\partial y \sim Kk$. It follows that unless $(\partial y/\partial\mathbf{x})_0$ happens to be the zero row vector $\mathbf{0}^T$, the objective can be made smaller (or larger) in the neighborhood of $\mathbf{x}_0$ simply by making k negative (or positive). By definition of an optimum, such a point cannot be optimal because there are feasible points nearby which are better. This reasoning gives the following well-known condition.

Necessary Condition for an Unconstrained Optimum. *If $\mathbf{x}^\star$ is an unconstrained optimum, then $(\partial y/\partial\mathbf{x})^\star = \mathbf{0}^T$.*

At any locally unconstrained point $\mathbf{x}_0$ where $(\partial y/\partial\mathbf{x})_0 \neq \mathbf{0}^T$, the objective y can always be improved, provided k is small enough to keep the remainder $0(\partial x_n^2)$ negligible. Taking the finite step $\Delta\mathbf{x} \equiv k(\partial y/\partial\mathbf{x})_0$, with $k < 0$ for minimization and $k > 0$ for maximization, is called the *gradient method*, since $(\partial y/\partial\mathbf{x})_0^T$ is the gradient of y.

Many optimization techniques involve finding points where the gradient vector vanishes. The trouble is that not every such point, which is called a *stationary point*, is the optimum. In fact, any unconstrained pessimum is also a stationary point. Moreover, there are *local optima* where no improvement is possible in the neighborhood, although large excursions will find better points. Local unconstrained optima are also stationary points, as are *saddles*, where some directions give local improvement while others make things worse. Distinguishing the optimum, called the *global optimum* when there are several local optima, from among all the stationary points requires study of the second-order terms in $0(\partial x_n^2)$, the remainder neglected in the first-order analysis. This task will be deferred until the first-order results have been applied to the fertilizer plant problem. First see if you can read the constrained contour map of Fig. 5-1 and distinguish among the various kinds of point.

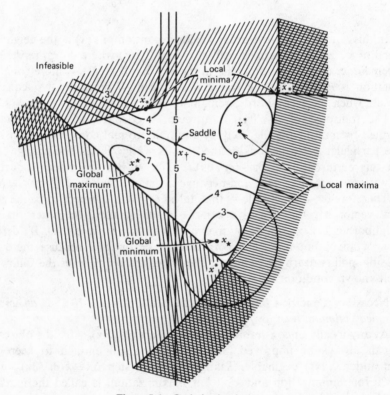

Figure 5-1 Optimization jargon

5.4. SCALING TO A STATIONARY POINT

Finding the maximum variable profit in the fertilizer plant problem will require choosing a base case, that is, a raw material rate m and a pressure p, and then improving it. This section shows how to generate a base case already near a stationary point.

At a stationary point, the first semilog derivative must vanish. Let $t_1, \ldots, t_5$ be the absolute values of the objective terms. That is,

$$b \equiv t_1 - t_2 - t_3 - t_4 - t_5$$

with $t_1 \equiv 360 m^{0.80}$; $t_2 \equiv 6.27(10^{-6}) m^{0.80} p^{2.30}$; $t_3 \equiv 9.92(10^3) m^{0.90} p^{-0.90}$; $t_4 \equiv 15.5(10^{-3}) mp$; $t_5 \equiv 0.650 m$. The semilog derivatives are written immediately as functions of the terms and the exponents. Notice that the negative signs are carried into the semilog derivatives. The subscript † (dagger) indicates that the derivatives are evaluated at stationary points.

$$\left(\frac{\partial b}{\partial \ln m} \right)_\dagger = (0.80 t_1 - 0.80 t_2 - 0.90 t_3 - t_4 - t_5)_\dagger = 0$$

$$\left(\frac{\partial b}{\partial \ln p} \right)_\dagger = (-2.30 t_2 + 0.90 t_3 - t_4)_\dagger = 0$$

It is reasonable to select a base case where these equations are satisfied, but there are too many terms to determine a unique solution. But if t_2 and t_5 are temporarily neglected, then the equations have the solution $t_4 = 0.90 t_3$ and $t_1 = 2.25 t_3$. In terms of m and p, this gives

$$t_4 = 15.5(10^{-3}) mp = 0.90(9.92)(10^3) m^{0.90} p^{-0.90} = 0.90 t_3$$

$$t_1 = 360 m^{0.80} = 2.25(9.92)(10^3) m^{0.90} p^{-0.90} = 2.25 t_3$$

This yields two nonlinear equations in the two unknown variables m and p. Linear equations are obtained by taking logarithms.

$$0.10 \log m + 1.90 \log p = 5.76042$$
$$-0.10 \log m + 0.90 \log p = 1.79239$$

The values of $\log m$ and $\log p$ are easily found on a pocket calculator

$$m = 10^{6.35} = 2.25(10^6) \qquad p = 10^{2.70} = 498$$

Then

$$(10^{-6}) b = 43.5 - 1.210 - 19.32 - 17.37 - 1.462 = 4.10$$

The second and fifth terms are indeed small, although not zero.

Define now dimensionless variables $x_1 \equiv m/(2.25)(10^6)$; $x_2 \equiv p/498$. These variables will be unity at the base case, and the coefficients in the new expression will be the values of the terms at the base case.

$$(10^{-6})b = 43.5x_1^{0.80} - 1.210x_1^{0.80}x_2^{2.30} - 19.32x_1^{0.90}x_2^{-0.90} - 17.37x_1x_2 - 1.462x_1$$

Scaling the objective in this way brings all the coefficients to the same order of magnitude, which not only eases later numerical work, but also gives the engineer a rough check for catching any errors that might have crept into the model. Even more convenient is to scale the objective to unity at the base case by dividing by the total. Then $y \equiv b/(4.10)(10^6)$.

$$y = 10.60x_1^{0.80} - 0.295x_1^{0.80}x_2^{2.30} - 4.71x_1^{0.90}x_2^{-0.90} - 4.24x_1x_2 - 0.357x_1$$

The coefficients in this expression represent multiples of the variable net profit at the base case. For example, the coefficient 10.60 on the sales term t_1 means that the gross sales are 10.60 times the profit in the base case. Similarly one can see that the reactor and energy costs are respectively, 4.71 and 4.24 times the base profit, while the compressor and raw material costs are only 29.5% and 35.7% of the base profit respectively.

In practice, the designer may have a particular base case in mind, say the present conditions for an existing installation. To avoid errors and to gain insight into the way the profit is distributed, scaling of this kind is strongly recommended.

5.5. UPPER-BOUNDING FUNCTION

It is instructive to regard the scaling procedure as locating the unique stationary point of an upper-bounding function. Let $u(\mathbf{x})$, the function obtained by neglecting the compressor and raw material costs, be called the *upper-bounding function* (ubf) for y, the scaled net profit to be maximized.

$$u(\mathbf{x}) \equiv 10.60x_1^{0.80} - 4.71x_1^{0.90}x_2^{-0.90} - 4.24x_1x_2$$
$$= y(\mathbf{x}) + 0.295x_1^{0.80}x_2^{2.30} + 0.357x_1 > y(\mathbf{x})$$

At the base case $(\mathbf{x})_0 \equiv (1,1)^T$, the scaled objective is unity, and so

$$u_0 = y_0 + 0.295 + 0.357 = 1.652$$

This base case was constructed to be at a stationary point of the ubf $u(\mathbf{x})$. Since the optimality conditions for $u(\mathbf{x})$ had a unique solution, this is known to be the only possible stationary point. The number of terms in $u(\mathbf{x})$ was critical in guaranteeing this uniqueness. As for posynomials, the number of terms was exactly one more than the number of variables, leaving zero degrees of difficulty.

If $(\mathbf{x})_0$ were a maximizer for $u(\mathbf{x})$, then $u((\mathbf{x})_0)$ would be a valid upper bound on $y(\mathbf{x})$. But at this juncture it is not known what kind of stationary point occurs at $(\mathbf{x})_0$. The value of $u(\mathbf{x})$ there is consequently indicated by $u_\dagger$.

$$u((\mathbf{x})_0) = u_\dagger$$

Notice that although clearly $y((\mathbf{x})_0) < u_\dagger$, it would be improper to write $y(\mathbf{x}) < u_\dagger$, since in principle there could be designs $\mathbf{x}$ better than at $(x)_0$ if $u_\dagger$ were a saddle or a minimum rather than a maximum for $u(\mathbf{x})$. Methods for testing for maximality will be presented in the next section. But rather than use them on $(\mathbf{x})_0$, where the gap between $y((\mathbf{x})_0)$ and $u_\dagger$ is still 65% of the value of y_0, let us construct a sharper ubf.

What is desired is an ubf with exactly three terms so that its stationary point will be unique. The strategy employed in constructing $u(x)$ was to neglect two of the five terms of $y(x)$. This time the two small terms will be combined with two large terms of the same sign to give a three-term upper bound.

Consider first the compressor cost $0.295 x_1^{0.80} x_2^{2.30}$ and the reactor cost $4.71 x_1^{0.90} x_2^{-0.90}$, which total 5.005 at $(\mathbf{x})_0$. The fractions of the compressor plus reactor cost are $0.295/5.005 = 0.0589$ and $4.71/5.005 = 0.941$, respectively. These fractions may be used as weights in an arithmetic-geometric mean inequality.

$$0.0589\left(\frac{0.295 x_1^{0.80} x_2^{2.30}}{0.295/5.005}\right) + 0.941\left(\frac{4.71 x_1^{0.90} x_2^{-0.90}}{4.71/5.005}\right)$$

$$\geq \left(\frac{0.295 x_1^{0.80} x_2^{2.30}}{0.295/5.005}\right)^{0.0589}\left(\frac{4.71 x_1^{0.90} x_2^{-0.90}}{4.71/5.005}\right)^{0.941}$$

$$= 5.005 x_1^{0.894} x_2^{-0.711}$$

with equality only at $(x)_0$. A similar inequality can be constructed for the energy plus raw material cost:

$$4.24 x_1 x_2 + 0.357 x_1 \geq 4.50 x_1 x_2^{0.922}$$

with equality only at $(\mathbf{x})_0$. This procedure of bounding a sum of several terms below by a single term is called *condensation*.

The above inequalities are reversed when both sides are multiplied by -1, giving *upper* bounds on the *negative* costs, which, when added to the gross profit, yields the original objective. Therefore

$$y(\mathbf{x}) \le 10.60 x_1^{0.80} - 5.005 x_1^{0.894} x_2^{-0.711} - 4.60 x_1 x_2^{0.922} \equiv u'(\mathbf{x})$$

This new upper-bounding function $u'(\mathbf{x})$ has only three terms as required, so it will have no more than one stationary point.

The optimality conditions for $u'(\mathbf{x})$ are:

$$x_1: \quad 0.80 u_1 - 0.894 u_2 - u_3 \quad = 0$$
$$x_2: \quad\quad\quad 0.711 u_2 - 0.922 u_3 = 0$$

where u_1, u_2, and u_3 are the terms of the upper-bounding function $u'(\mathbf{x})$. The coordinates of $(\mathbf{x})_1$, the stationary point for $u'(\mathbf{x})$, are $(\mathbf{x})_1 = (0.656, 0.923)^T$. At this stationary point, the objective and upper-bounding functions are

$$y[(\mathbf{x})_1] = 7.657 - 0.175 - 3.465 - 2.568 - 0.234 = 1.125$$
$$u'[(\mathbf{x})_1] = 7.567 - 3.636 - 2.804 = 1.128$$

The difference is small enough to justify seeing if $(\mathbf{x})_1$ is a maximum for $u'(\mathbf{x})$. This will involve studying the curvature of $u'(\mathbf{x})$ at its stationary point.

5.6. CURVATURE AT A STATIONARY POINT

By definition, the first-order part of the Taylor expansion for the objective vanishes at an unconstrained stationary point $\mathbf{x}_+$

$$(\partial y)_+ \sim \left(\frac{\partial y}{\partial \mathbf{x}}\right)_+ \partial \mathbf{x} = \mathbf{0}^T \partial \mathbf{x} = 0$$

Thus distinguishing between the various kinds of stationary point requires examining the second-order part of the Taylor expansion.

At any point $\mathbf{x}$ the Taylor expansion to second order is

$$\partial y = \left(\frac{\partial y}{\partial \mathbf{x}}\right) \partial \mathbf{x} + \frac{1}{2} \sum_{m=1}^{N} \sum_{n=1}^{N} \left(\frac{\partial^2 y}{\partial x_m \partial x_n}\right) \partial x_m \partial x_n + 0(\partial x_n^3)$$

where $0(x_n^3)$ represents the remainder of degree three and higher in ∂x_n. Henceforth the symbol $\approx$ will represent the first- and second-order parts of the Taylor expansion with the third-order remainder omitted. At a stationary point $x_\dagger$, the first-order part vanishes, so

$$(\partial y)_\dagger \approx \tfrac{1}{2} \sum \sum \left(\frac{\partial^2 y}{\partial x_m \partial x_n} \right)_\dagger \partial x_m \partial x_n.$$

In the example, the point x_1 is stationary for the upper-bounding function $u'(x)$ wrt the logarithmic independent variables $\ln x_1$ and $\ln x_2$, because $(\partial u'/\partial \ln x)_\dagger = (0,0) \equiv 0^T$. Here the symbol $\ln x$ is the column vector of natural logarithms of $x_1, \ldots, x_N$.

$$\ln x \equiv (\ln x_1, \ldots, \ln x_N)^T$$

Let the *second* logarithmic derivatives be arranged in an N^2 symmetric matrix, symbolized by $\partial^2 u'/(\partial \ln x)^2$.

$$\frac{\partial^2 u'}{(\partial \ln x)^2} \equiv \begin{bmatrix} \dfrac{\partial^2 u'}{\partial(\ln x_1)^2} & \cdots & \dfrac{\partial^2 u'}{\partial \ln x_1 \partial \ln x_N} \\ \vdots & & \vdots \\ \dfrac{\partial^2 u'}{\partial \ln x_N \partial \ln x_1} & \cdots & \dfrac{\partial^2 u'}{(\partial \ln x_N)^2} \end{bmatrix}$$

In the example,

$$\frac{\partial^2 u'}{\partial(\ln x)^2} = \frac{\partial^2 u_1}{\partial(\ln x)^2} - \frac{\partial^2 u_2}{\partial(\ln x)^2} - \frac{\partial^2 u_3}{\partial(\ln x)^2}$$

where u_1, u_2, and u_3 are the terms of $u(x)$.

Each of these second derivative matrices is easily written as a matrix multiple of the corresponding term. Thus

$$\frac{\partial^2 u_1}{\partial(\ln x)^2} = \begin{pmatrix} (0.80)^2 & 0 \\ 0 & 0 \end{pmatrix} u_1 = \begin{pmatrix} 0.64 & 0 \\ 0 & 0 \end{pmatrix} u_1$$

$$\frac{\partial^2 u_2}{\partial(\ln x)^2} = \begin{bmatrix} (0.894)^2 & (0.894)(-0.711) \\ (0.894)(-0.711) & (-0.711)^2 \end{bmatrix} u_2$$

$$= \begin{pmatrix} 0.799 & -0.636 \\ -0.636 & 0.506 \end{pmatrix} u_2$$

$$\frac{\partial^2 u_3}{\partial(\ln x)^2} = \begin{bmatrix} 1^2 & 1(0.922) \\ 1(0.922) & (0.922)^2 \end{bmatrix} u_3 = \begin{pmatrix} 1 & 0.922 \\ 0.922 & 0.850 \end{pmatrix} u_3$$

This follows from the general relation, a very important property of power functions, that when $u_t = C_t \prod x_n^{\alpha_{tn}}$,

$$\frac{\partial^2 u_t}{\partial \ln x_m \partial \ln x_n} = \frac{\partial}{\partial \ln x_m}\left(\frac{\partial u_t}{\partial \ln x_n}\right) = \alpha_{tn}\frac{\partial u_t}{\partial \ln x_m} = \alpha_{tm}\alpha_{tn}u_t$$

At the stationary point x_1 then

$$\left(\frac{\partial^2 u'}{\partial(\ln x)^2}\right)_1 = \begin{pmatrix} 0.64 & 0 \\ & 0 \end{pmatrix} 7.57 - \begin{pmatrix} 0.799 & -0.636 \\ & 0.506 \end{pmatrix} 3.46 - \begin{pmatrix} 1 & 0.922 \\ & 1 \end{pmatrix} 2.80$$

$$= \begin{pmatrix} -0.730 & -0.382 \\ & -4.56 \end{pmatrix}$$

where, to save print, the elements below the diagonal are omitted in these symmeric matrices. The second-order Taylor expansion in the neighborhood of x_1 is therefore

$$(\partial u')_1 \approx \tfrac{1}{2}(\partial \ln x_1, \partial \ln x_2)\begin{pmatrix} -0.730 & -0.382 \\ & -4.56 \end{pmatrix}\begin{pmatrix} \partial \ln x_1 \\ \partial \ln x_2 \end{pmatrix}$$

$$= \tfrac{1}{2}\left[-0.730(\partial \ln x_1)^2 + 2(-0.382)\partial \ln x_1 \partial \ln x_2 - 4.56(\partial \ln x_2)^2 \right]$$

This expression is an example of what mathematicians call a *homogeneous quadratic form*, every term being of second degree.

The nature of the stationary point can always be determined by completing the square to obtain a quadratic form free of cross-product terms $\partial \ln x_m \partial \ln x_n$. In the example,

$$-0.730(\partial \ln x_1)^2 + 2(-0.382)\partial \ln x_1 \partial \ln x_2 - 4.56(\partial \ln x_2)^2$$

$$= -0.730(\partial \ln x_1)^2 + 2(-0.382)\partial \ln x_1 \partial \ln x_2 - \frac{(0.382)^2}{0.730}(\partial \ln x_2)^2$$

$$+ \frac{(0.382)^2}{0.730}(\partial \ln x_2)^2 - 4.56(\partial \ln x_2)^2$$

$$= -(0.854\partial \ln x_1 - 0.447\partial \ln x_2)^2 - 4.36(\partial \ln x_2)^2$$

This quadratic form can never be positive, since each term is the negative of a perfect square. Moreover, it will be strictly negative everywhere in the neighborhood of x_1, for by definition this neighborhood is such that $\partial \ln x \neq 0$. Hence to second-order the objective u' achieves its greatest value within the neighborhood of x_1 at x_1 itself: $(\partial u')_1 < 0$ for all $\partial \ln x \neq 0$. This in fact satisfies the definition of an *unconstrained local maximizer*.

In general, an *unconstrained local maximizer* x^* is any point where $\partial y < 0$ in every neighborhood. The value of the objective there, called a local maximum, is symbolized by y^*. An *unconstrained local minimizer* x_* and *local minimum* y_* are defined similarly, except that $y > 0$ in every neighborhood of a local minimizer. Any stationary point with both better and worse points in its neighborhood is called a *saddle*.

This analysis of the matrix of second partial derivatives (called the *hessian*) has proven that x_1 is a local maximizer for the ubf $u'(x)$. This is an application of the important condition:

Sufficient Condition for an Unconstrained Local Maximum (Minimum).
At a stationary point $x_\dagger$, *let* $\partial x^T (\partial^2 y / \partial x^2)_\dagger \partial x < 0 \, (> 0)$ *for all* $\partial x \neq 0$. *Then* $x_\dagger = x^* \, (= x_*)$, *that is, it is a local maximizer (minimizer).*

5.7. A SATISFACTORY DESIGN

In the chemical plant problem, the design x_1 has been shown to produce a local maximum for the ubf $u'(x)$, so $u'(x_1) = (u')^*$. Moreover, the ubf cannot have any other stationary points, which means that x_1 is the global, not just the local, maximizer for $u'(x)$: $(u')^* = (u')^*$. Hence $(u')^*$ is an upper bound for the profit $y(x)$.

$$y(x) \leq u'(x) \leq (u')^*$$

In particular, the design x_1 is such that $y(\mathbf{x}_1) = 1.125 \le (u')^\star = 1.128$. The global maximum for y must be contained in the same interval.

$$1.125 = y(\mathbf{x}_1) \le y^\star \le (u')^\star = 1.128$$

Equivalently, $(u')^\star - y(\mathbf{x}_1) \le 0.003$, an interval well within the three figure accuracy of the original coefficients, amounting to a maximum gap of 0.01 (1%). The design $\mathbf{x}_1$ is just one of many having a profit within 1% of the theoretical global maximum and therefore indistinguishable from it (see Fig. 5-2). All the engineer wants is any one of these designs, any of which is satisfactory in the sense of Sec. 3.7.

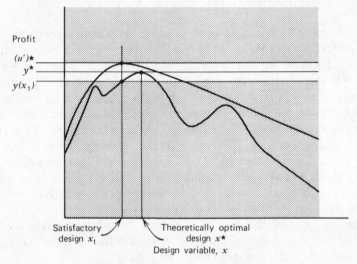

Figure 5-2   Satis quod sufficit

From now on whenever a search is ended because a satisfactory design has been generated, the abbreviation *sqs*, standing for *satis quod sufficit* (enough is as good as a feast) will follow the the satisfactory design. This is the designers' equivalent of the *qed*, for *quod erat demonstrandum* (which was to be proven), used by mathematicians at the end of a proof.

5.8.   SIGNOMIALS

Sums having at least one negative power function term mixed in with the positive ones were at first called "generalized polynomials," but Duffin's shorter word "signomial" is more widely used today. Although pronounced sig-*nom*-i-al, the key root is the word "sign," the first four letters, suggesting that the function is like a *posy*nomial except that its terms have mixed signs. These signs are represented by *signum* functions (pronounced *signum*) σ_t having values either plus or minus one, the index t identifying the term. Thus the terms t_t $(t = 1, \ldots, T)$ of a signomial function are defined just as for a posynomial: $t_t(\mathbf{x}) = C_t \Pi_{n=1}^{N} x_n^{\alpha_{tn}}$ with $C_t > 0$ and α_{tn} real, but the signomial function $s(\mathbf{x})$ has signum functions multiplying each term to account for the sign.

$$s(\mathbf{x}) = \prod_{t=1}^{T} \sigma_t t_t(\mathbf{x})$$

In the preceding example, $\sigma_1 = 1$ and $\sigma_2 = \sigma_3 = \sigma_4 = \sigma_5 = -1$ by definition. Like the coefficients C_t and exponents α_{tn}, the signum functions are given as part of the problem definition.

The first semilog derivatives are

$$\frac{\partial s(\mathbf{x})}{\partial \ln x_n} = \sum_{t=1}^{T} \sigma_t \alpha_{tn} t_t(\mathbf{x}) \qquad n = 1, \ldots, N$$

At a stationary point $\mathbf{x}_\dagger$ these are by definition zero:

$$\sum_{t=1}^{T} \sigma_t \alpha_{tn} t_t(\mathbf{x}_\dagger) = 0 \qquad n = 1, \ldots, N$$

Division by the absolute value of $s(\mathbf{x})$ gives expressions in terms of positive *primal weights* $w_t \equiv t_t / |s|$:

$$\sum_{t=1}^{T} \sigma_t \alpha_{tn} (w_t)_\dagger = 0 \qquad n = 1, \ldots, N$$

These primal weights must also satisfy a normality condition obtained by dividing the signomial by its absolute value:

$$\frac{s}{|s|} = \sum_{t=1}^{T} \sigma_t w_t = \sigma_0$$

where σ_0 is the sign of s, which could be either positive or negative, although in this analysis it is assumed not to vanish ($s \neq 0$).

As in posynomial geometric programming, let these linear equations be written as functions of *dual* variables δ_t, one for each term, which do not necessarily correspond to the primal weights.

$$\sum_{t=1}^{T} \sigma_t \alpha_{tn} \delta_t = 0 \qquad n = 1, \ldots, N$$

$$\sum_{t=1}^{T} \sigma_t \delta_t = \sigma_0$$

For posynomials, where $\sigma_t = 1$ for all $t = 0, 1, \ldots, T$, these would be identical with the orthogonality and normality conditions of geometric programming. Notice, however, that the geometric inequality was not used to derive them, which is why they remain valid even for signomials while the orthogonality conditions do not. To emphasize this subtle but important distinction, these more general equations will be called, respectively, the *stationarity* and *s-normality* conditions, the "s," of course, being an abbreviation for "signomial."

The concept of degrees of difficulty is the same as it was for posynomials. Thus when $T = N + 1$, the stationarity and s-normality conditions have a unique solution $\boldsymbol{\delta}^{\dagger}$, which must match the primal weights $\mathbf{w}_{\dagger}$ at the stationary point, since both $\boldsymbol{\delta}^{\dagger}$ and $\mathbf{w}_{\dagger}$ satisfy the same equations. This also proves that the stationary point is unique when there are no degrees of difficulty. Equating $\boldsymbol{\delta}^{\dagger}$ to $\mathbf{w}_{\dagger}$ in fact gives T relations between the N design variables—the *invariance conditions*.

$$\delta_t^{\dagger} = (w_t)_{\dagger} = C_t \prod_{n=1}^{N} \frac{x_{n_{\dagger}}^{\alpha_{tn}}}{|s|_{\dagger}} \qquad t = 1, \ldots, T$$

Taking common logarithms gives a consistent set of linear equations in the logarithms of the unknown design variables $\mathbf{x}_{\dagger}$ at the stationary point

$$\sum_{n=1}^{N} \alpha_{tn} \log(x_n)_{\dagger} = \log \frac{\delta_t^{\dagger} |s|_{\dagger}}{C_t} \qquad t = 1, \ldots, T$$

These can be solved once $|s|_{\dagger}$ has been found, as follows.

In the example, the design $\mathbf{x}_{\dagger}$ was found without calculating $|s|_{\dagger}$ by working with terms rather than weights. Yet $|s|_{\dagger}$ is easily computed by the formula

$$|s|_{\dagger} = \sigma_0 s_{\dagger} = |s|_{\dagger}^{\sigma_0 \Sigma \sigma_t \delta_t^{\dagger}} = \left[\prod_{t=1}^{T} |s|_{\dagger}^{\sigma_t \delta_t^{\dagger}} \right]^{\sigma_0} = \left[\prod_{t=1}^{T} \left(\frac{t_t}{\delta_t^{\dagger}} \right)^{\sigma_t \delta_t^{\dagger}} \right]^{\sigma_0}$$

$$= \left[\prod_{t=1}^{T} \left(\frac{C_t}{\delta_t^{\dagger}} \right)^{\sigma_t \delta_t^{\dagger}} \right]^{\sigma_0}$$

The design variables x cancel out of this expression because the dual variables $\pmb{\delta}^{\dagger}$ satisfy the stationarity conditions by definition.

This formula could have been used in Sec. 5.5 on the zero-degree-of-difficulty upper-bounding function $u'(\mathbf{x}) \equiv 10.60 x_1^{0.80} - 5.005 x_1^{0.894} x_2^{-0.711} - 4.60 x_1 x^{0.922}$ to find its value at its unique stationary point, without finding the corresponding design first as was done there. The stationarity and s-normality conditions are

$$
\begin{array}{lll}
x_1: & 0.80\delta_1^{\dagger} - 0.894\delta_2^{\dagger} & -\delta_3^{\dagger} = 0 \\
x_2: & 0.711\delta_2^{\dagger} - 0.922\delta_3^{\dagger} = 0 \\
& \delta_1^{\dagger} \quad -\delta_2^{\dagger} \quad -\delta_3^{\dagger} = 1
\end{array}
$$

The solution is $\pmb{\delta}^{\dagger} = (6.71, 3.22, 2.49)$, and so the value of $(u')_{\dagger}$ is

$$
(u')_{\dagger} = (10.60/6.71)^{6.71}(5.005/3.22)^{-3.22}(4.60/2.43)^{-2.49} = 1.128
$$

as computed by the other method in Sec. 5.5.

The reader may be puzzled that σ_0 was set to plus one in this example, for it is not really known in advance that u' is positive at its stationary point. This was in fact a guess, based on the positivity of the profit function $y(\mathbf{x})$ at the base case. But it would have been immediately apparent if this hunch had been wrong, for then all of the dual variables would have been negative, violating their definition. Then changing σ_0 to -1 would make all their signs positive.

Just as for posynomials, obtaining mixed signs for the dual variables in a zero-degree-of-difficulty problem would mean that the objective had no stationary point. This would of course make it unbounded and force reexamination of the model.

The relations derived here hold only at the stationary point $\mathbf{x}_{\dagger}$. But the only time it is easy to find the values of the signomial-dual variables $\pmb{\delta}^{\dagger}$ at $\mathbf{x}_{\dagger}$ is when there are zero degrees of difficulty, for then the stationarity and s-normality conditions have a unique solution. One method for solving signomial problems then requires condensation to zero degrees of difficulty and finding $\mathbf{x}_{\dagger}$ for the condensed problem from the invariance conditions. If substitution of $x_{\dagger}$ into the original objective gives a value matching that for the condensation, then the stationary point has been found. Proof that this point is an optimum requires checking the matrix $(\partial^2 s / \partial (\ln \mathbf{x})^2)_{\dagger}$ as described in Sec. 5.6. In the present notation,

$$
\frac{\partial^2 s}{\partial (\ln x_i) \partial (\ln x_j)} = \sum_{t=1}^{T} \sigma_t \alpha_{ti} \alpha_{tj} t_t
$$

This matrix will be positive definite at a minimum and negative definite at a maximum, as in the example.

This condensation procedure may not always be applicable. Section 5.9 describes a partial invariance scheme for problems with degrees of difficulty, and Sec. 5.10 develops direct search methods using the easily computed first and second semilog derivatives of signomials.

5.9. SIGNOMIAL PARTIAL INVARIANCE

It may not always be clear how to condense a signomial problem to zero degrees of difficulty, or one may prefer guessing dual variables to guessing weights. In such circumstances, a number of dominant terms equal to the number of independent variables plus one can be selected just as for posynomials. The stationarity and s-normality conditions are then solved for the dual variables corresponding to these dominant terms. They will be linear functions of the remaining dual variables, designated "recessive" for reference. Fixing the recessives then determines a set of dual variables satisfying the stationarity and s-normality conditions, although these will usually *not* be the desired values $\delta^\dagger$ corresponding to a stationary point $x_\dagger$ of the objective $s(x)$.

Nevertheless the δ can be used to evaluate the following function, which by analogy with the dual function of posynomial geometric programming will be called the *signomial dual function* (s-dual, for short)

$$\bar{d}(\delta) = \left[\prod_{t=1}^{T} \left(\frac{C_t}{\delta_t} \right)^{\sigma_t \delta_t} \right]^{\sigma_0}$$

As derived in the preceding section, this s-dual function will equal the signomial objective at, and only at, a stationary point. Anywhere else the invariance conditions using $\bar{d}$ instead of s will be inconsistent. Yet a design x can be generated by any consistent subset. The value of the objective for this design can then be compared to $\bar{d}$ to see if further iteration is needed. Unlike the posynomial situation, in which the dual function is always a lower bound on the objective, the s-dual function cannot be used as either an upper or a lower bound without second-order analysis. This is because the dual relations for posynomials were derived from the geometric inequality, whereas those for signomials stemmed from the necessary conditions for a stationary point and are only valid there.

The example following will show how to apply the partial invariance method to the fertilizer plant profit problem. Generation of the base case could be considered already an application of this procedure in which the recessive duals were set to zero. As a clearer illustration, consider the iteration from the base case, where the scaled signomial objective, expressed in terms of the scaled variables x_1 and x_2, is

$$y = 10.6x_1^{0.8} - 0.295x_1^{0.8}x_2^{2.3} - 4.71x_1^{0.9}x_2^{-0.9} - 4.24x_1x_2 - 0.357x_1$$

The stationarity and s-normality conditions are

$$
\begin{aligned}
x_1: &\quad 0.8\delta_1 - 0.8\delta_2 - 0.9\delta_3 - \delta_4 - \delta_5 = 0 \\
x_2: &\quad -2.3\delta_2 + 0.9\delta_3 - \delta_4 \phantom{{}-\delta_5} = 0 \\
&\quad \delta_1 - \delta_2 - \delta_3 - \delta_4 - \delta_5 = 1
\end{aligned}
$$

Solution for the dominant duals δ_1, δ_3, and δ_4 gives

$$\delta_1 = 6.428 + 1.822\delta_2 - 0.357\delta_5$$
$$\delta_3 = 2.857 + 1.643\delta_2 - 0.714\delta_5$$
$$\delta_4 = 2.571 - 0.821\delta_2 - 0.643\delta_5$$

Let the recessive dual variables assume values equal to their weights at the base case, namely, $\delta_2 = 0.295$ and $\delta_5 = 0.357$. Then $\delta_1 = 6.838$, $\delta_3 = 3.087$, and $\delta_4 = 2.099$. The s-dual function for these values is

$$d_s = \left(\frac{10.6}{6.838}\right)^{6.838}\left(\frac{0.295}{0.295}\right)^{-0.295}\left(\frac{4.71}{3.087}\right)^{-3.087}$$

$$\times \left(\frac{4.24}{2.099}\right)^{-2.099}\left(\frac{0.357}{0.357}\right)^{-0.357} = 1.243$$

The invariance conditions for the two largest terms will be used to generate a design. The first term gives $10.6x_1^{0.8} = 6.838(1.243)$, or $x_1 = 0.759$. The third similarly gives $x_2 = 0.953$, for which the scaled profit is 1.114. This is 13% below the s-dual function, which is better than the 65% gap at the base case but not as good as the 0.3% spread obtained by condensation. As for the gravel sled of Chapter 3, condensation performed better than partial invariance, which should be used only when degrees of difficulty are unavoidable due to insufficient positive (negative) terms in a minimization (maximization) problem.

5.10. DIRECT SEARCH

Although not of particular interest here, it can be proven that the s-dual variables $\boldsymbol{\delta}^\dagger$ corresponding to $\mathbf{x}_\dagger$, the stationary point of a signomial $s(\mathbf{x})$, are also at a stationary point of the s-dual function $\bar{d}(\boldsymbol{\delta})$. This is not very useful to designers, however, because it is not known whether the s-dual stationary point is minimum, maximum, or neither. If one is to search directly for a stationary point, it makes sense to look for the primal one because the second-order character of the desired design is specified in the problem statement.

Among the myriad methods for seeking an optimum by direct manipulation of the design variables, the venerable Newton-Raphson method is best suited for signomial problems because it uses second and first derivatives, both easily available in semilog form. After a quick derivation, this method will be demonstrated on the fertilizer plant profit problem.

Consider the Taylor expansion of the column vector of first semilog derivatives of a signomial $s(\mathbf{x})$ in the neighborhood of a base case $\mathbf{x}_0$.

$$\left(\frac{\partial y}{\partial \ln \mathbf{x}}\right)^T \sim \left(\frac{\partial y}{\partial \ln \mathbf{x}}\right)_0 + \left(\frac{\partial^2 y}{\partial (\ln \mathbf{x})^2}\right)_0 \partial \ln \mathbf{x}$$

The N-R method, as applied to finding a stationary point, involves replacing the infinitesimal $\partial \ln \mathbf{x}$ by a finite change $\Delta \ln \mathbf{x}$ and setting the left member to the null vector $\mathbf{0}$, the value at the desired stationary point $\mathbf{x}_+$. Rearrangement gives $\Delta \ln \mathbf{x}$ implicitly as the solution of the N^2 system of linear equations whose matrix form is

$$\left(\frac{\partial^2 y}{\partial (\ln \mathbf{x})^2}\right)_0 \Delta \ln x = -\left(\frac{\partial y}{\partial \ln \mathbf{x}}\right)_0^T$$

In the fertilizer plant profit example of the preceding section, the derivative formulas of Sec. 5.8 give, at the base case $x_0 = (1, 1)^T$,

$$\left(\frac{\partial y}{\partial \ln x_1}\right)_0 = 0.8(10.6) - 0.8(0.295) - 0.9(4.71) - 4.24 - 0.357 = -0.592$$

$$\left(\frac{\partial y}{\partial \ln x_2}\right)_0 = -2.3(0.295) + 0.9(4.71) - 4.24 = -0.680$$

$$\left[\frac{\partial^2 y}{\partial (\ln x_1)^2}\right]_0 = (0.8)^2 10.6 - (0.8)^2 0.295 - (0.9)^2 4.71 - 4.24 - 0.357 = -1.817$$

$$\left[\frac{\partial^2 y}{\partial (\ln x_1) \partial (\ln x_2)}\right]_0 = -(0.8)(2.3)0.295 - (0.9)^2 4.71 - 4.24 = -0.968$$

$$\left[\frac{\partial^2 y}{\partial (\ln x_2)^2}\right]_0 = -(2.3)^2 0.295 - (0.9)^2 4.71 - 4.24 = -9.616$$

The desired changes $\Delta\ln\mathbf{x}$ are the unique solutions to the linear equations

$$-1.817\Delta\ln x_1 - 0.968\Delta\ln x_2 = -(-0.592)$$
$$-0.968\Delta\ln x_1 - 9.616\Delta\ln x_2 = -(-0.680)$$

These are $\Delta\ln x_1 = -0.304$ and $\Delta\ln x_2 = -0.0401$, for which the coordinates of the new point are $(0.737, 0.961)$. This gives a design worth 1.118, only 0.7% below the design obtained by condensation. This is in fact just inside the 1% satisfaction threshold, but this would not be known if the ubf had not been constructed. More iterations would be needed unless an upper bound is computed by one of the methods given earlier in the chapter.

The N-R method is then a back-up technique for use when the other procedures cannot be used. It can in fact be used to achieve early improvement of a bad start, the other methods being employed to generate bounds for terminating the search when improvement slows.

The N-R method does not always work this well. Sometimes speed of convergence is offset by instability. Problems with the N-R method, together with suggestions for overcoming, or at least controlling them, are discussed in the next section. But first, the gradient method, mentioned briefly in Sec. 5.3, is applied to this problem so that its performance can be compared with that of the N-R method.

The gradient method is widely used in direct search because it does not require second derivatives as does the N-R method. It involves changing each coordinate in proportion to the numerical value of the first partial derivative at the point of departure, which in the example would give $x_1 = -0.592k$ and $x_2 = -0.680k$, where k is a search parameter generating the direction of investigation. For sufficiently small k, improvement in the objective is guaranteed, and one can quickly determine on a programmable calculator that the highest value achievable is $y = 1.059$ when k is between 0.14 and 0.16. Taking $k = 0.15$ gives a new point $\mathbf{x} = (0.911, 0.898)^T$. This only achieves about half the improvement that the N-R method accomplished with a comparable amount of effort. Iteration of the gradient method from the new design is left as an exercise. Despite its mediocre performance on this example, the gradient method is worth knowing because it will give improvement when the N-R method does not. Potential difficulties with this temperamental N-R method, and how to overcome them, are the business of the rest of the chapter.

5.11. PITFALLS

At first glance, the N-R method may appear clearly superior to the gradient method, since the former makes large, finite steps, contrasting with the relatively small steps of the latter. Yet the N-R method must be used cautiously for two reasons. First, its large step may carry it outside the region where the quadratic approximation is valid, causing oscillation or even divergence of the algorithm. The second reason is less obvious—the method may go in the wrong direction, causing the objective to get worse instead of better. This could happen when there are several stationary points, for the N-R method migrates toward the nearest, rather than the best, stationary point. But even when there is only one stationary point, and that the optimum, the method could still go in the wrong direction, as will now be demonstrated.

Consider the change $a(\Delta x)$ in the objective predicted by the approximation.

$$a(\Delta x) = (y_x)_0 \Delta x + \tfrac{1}{2} \Delta x^T (y_{xx})_0 \Delta x$$

But

$$(y_x)_0 = -\Delta x^T (y_{xx})_0$$

so

$$a(\Delta x) = -\tfrac{1}{2} \Delta x^T (y_{xx})_0 \Delta x$$

This is a quadratic form in Δx, and its sign is at present uncertain, although ways of relating it to properties of the matrix $(y_{xx})_0$ are developed in the section following. To emphasize this sign ambiguity, substitute $(y_x)_0$ for $-\Delta x^T (y_{xx})_0$ in $a(\Delta x)$ to obtain

$$a(\Delta x) = \tfrac{1}{2} (y_x)_0 \Delta x$$

For this inner product to be negative, as desired for minimization, the step direction must form an obtuse angle with the gradient, but nothing in the method guarantees this.

Hence one should always verify that the objective actually improves after a N-R step. If it does not, move the same distance in the opposite direction, i.e., make a step $-\Delta x$. For in this case, the predicted change in y is

$$a(-\Delta x) = (y_x)_0(-\Delta x) + \tfrac{1}{2}(-\Delta x^T)(y_{xx})_0(-\Delta x)$$

$$= -(y_x)_0 \Delta x + \tfrac{1}{2} \Delta x^T (y_{xx})_0 \Delta x = -\left[-\Delta x^T (y_{xx})_0 \right]\Delta x + \tfrac{1}{2}\Delta x^T(y_{xx})_0 \Delta x$$

$$= (3/2)\Delta x^T (y_{xx})_0 \Delta x = -3a(\Delta x)$$

Thus if the N-R move gave an increase in y, the same length step in the opposite direction will, if the approximation is accurate, give a decrease three times as great in absolute value as the N-R increase.

Another, milder, annoyance occurs when $(y_{xx})_0$ does not have full rank. In this case any component of Δx vanishing in the elimination of the N-R equations can be specified arbitrarily. In practice such components should be made small enough to be consistent with the validity of the approximation. The next section seeks to explain all these mysteries.

5.12. QUADRATIC FORMS

To explain and correct the difficulties of the N-R method described in the preceding section, the mathematics of quadratic forms must be developed in more detail than in Sec. 5.6. The results will then be used to distinguish local minima from among the other kinds of stationary point.

The second-order part $\partial x^T y_{xx} \partial x$ of the Taylor expansion of $y(x)$, which governs the behavior of the N-R method, is a quadratic form related to the local curvature of the objective function surface. But instead of this differential quadratic form, let us study the general quadratic form $Q(x) = x^T Q x$, where Q is the N^2 symmetric matrix of the quadratic form $Q(x)$. The components of x are unrestricted; they can be positive, negative, or zero, and $Q(0) = 0$. Consider the sign of $Q(x)$ when $x \neq 0$.

When $Q(x) > 0$ for all $x \neq 0$, the quadratic form and its matrix Q are said to be *positive definite*. For example, $x_1^2 + x_2^2$ is positive definite since each term is nonnegative, and the sum is positive unless $x = 0$. Here $Q = I$, the 2×2 unit matrix.

Similarly, when $Q(x) < 0$ for all $x \neq 0$, then $Q(x)$ and Q are called *negative definite*. An example is $Q(x) = -x_1^2 - x_2^2$, for which $Q = -I$.

A quadratic form which can take either sign is called *indefinite*, as for $Q(x) = x_1^2 - x_2^2$, which is positive for $|x_1| > |x_2|$, negative for $|x_1| < |x_2|$, and zero for $|x_1| = |x_2|$. Here

$$Q = \begin{pmatrix} 1 & 0 \\ 0 & -1 \end{pmatrix}$$

If $Q(x)$ can vanish without changing sign, it is called *semidefinite*. Thus if $Q(x) \geq 0$ (≤ 0), $Q(x)$ and Q are called *positive* (*negative*) *semidefinite*. If $x = (x_1, x_2)^T$, then x_1^2 is positive semidefinite and $-x_1^2$ is negative semidefinite. The corresponding matrices are

$$\begin{pmatrix} 1 & 0 \\ 0 & 0 \end{pmatrix} \quad \text{and} \quad \begin{pmatrix} -1 & 0 \\ 0 & 0 \end{pmatrix}$$

which, having rank one, are singular.

Since $\mathbf{x}$ is real, every quadratic form is in one of these classes. It is easy to tell which class whenever $\mathbf{Q}$ is diagonal, for then $Q(\mathbf{x})$ is a sum or difference of perfect squares, each necessarily positive. If all of these diagonal elements are positive (negative), $\mathbf{Q}$ is positive (negative) definite, although any zero on the diagonal in this case makes the form *semi*definite. If both signs appear, $\mathbf{Q}$ is indefinite.

However, Q may not be diagonal, as in the following example where

$$(y_{xx})_0 = \begin{pmatrix} 2 & 1 \\ 1 & 2 \end{pmatrix} \equiv \mathbf{Q} \quad \text{and} \quad Q(\mathbf{x}) = 2x_1^2 + 2x_1 x_2 + 2x_2^2$$

The character of $Q(\mathbf{x})$ is obscured by the cross-product term $2x_1 x_2$, because this can be either positive or negative according to whether the signs of x_1 and x_2 are the same or not. Since the coefficients of the perfect squares, or equivalently, the diagonal elements of the *non*diagonal matrix $\mathbf{Q}$, are all positive, the form can be positive for some values of $\mathbf{x}$. The question remains: can it be made negative or zero?

The simplest way to find out is to complete the square, for this gives a new quadratic form, equal to the old one, but with no cross products. In the example

$$Q(\mathbf{x}) = 2\left(x_1^2 + x_1 x_2 + x_2^2\right) = 2\left[\left(x_1^2 + x_1 x_2 + \tfrac{1}{4}x_2^2\right) + \left(x_2^2 - \tfrac{1}{4}x_2^2\right)\right]$$

$$= 2\left[\left(x_1 + \tfrac{1}{2}x_2\right)^2 + \tfrac{3}{4}x_2^2\right]$$

Since this is a sum of two perfect squares, $Q(\mathbf{x})$ is proven positive definite.

Completion of the square may be regarded as a linear transformation on $\mathbf{x}$. Thus let

$$\tilde{\mathbf{x}} \equiv \begin{pmatrix} \tilde{x}_1 \\ \tilde{x}_2 \end{pmatrix} \equiv \begin{pmatrix} x_1 + \tfrac{1}{2}x_2 \\ x_2 \end{pmatrix} = \begin{pmatrix} 1 & \tfrac{1}{2} \\ 0 & 1 \end{pmatrix} \begin{pmatrix} x_1 \\ x_2 \end{pmatrix} \equiv \mathbf{T}\mathbf{x}$$

where T is a *transformation matrix*. Then in the example,

$$Q(\mathbf{x}) = 2\left(\tilde{x}_1^2 + \tfrac{3}{4}\tilde{x}_2^2\right) = 2\left[\tilde{\mathbf{x}}^T \begin{pmatrix} 1 & 0 \\ 0 & \tfrac{3}{4} \end{pmatrix} \tilde{\mathbf{x}}\right] \equiv 2\tilde{\mathbf{x}}^T \mathbf{D}\tilde{\mathbf{x}} \equiv Q(\tilde{\mathbf{x}})$$

where the coefficients of the squares appear as elements of the *diagonal* matrix $\mathbf{D}$. Thus $Q(\mathbf{x}) = \mathbf{x}^T \mathbf{Q}\mathbf{x} = \tilde{\mathbf{x}}^T \mathbf{D}\tilde{\mathbf{x}} = Q(\tilde{\mathbf{x}})$. Since $\mathbf{D}$ is diagonal, its definiteness is immediate by inspection of its diagonal elements. Section 5.14 shows that positive definiteness makes the N-R method work effectively.

The process of finding a diagonal matrix $\mathbf{D}$ corresponding to $\mathbf{Q}$ is called *diagonalizing* $\mathbf{Q}$. In mathematical terminology $\mathbf{Q}$ and $\mathbf{D}$ are said to be *similar* when their quadratic forms are equal under the transformation $\mathbf{T}$, which is called a *similarity transformation*. These matrices are related by $\mathbf{Q} = \mathbf{T}^T\mathbf{D}\mathbf{T}$ because $\mathbf{x}^T\mathbf{D}\mathbf{x} = \mathbf{x}^T\mathbf{T}^T\mathbf{D}\mathbf{T}\mathbf{x} = \mathbf{x}^T\mathbf{Q}\mathbf{x}$.

An example of an indefinite matrix is

$$\begin{pmatrix} 1 & 2 \\ 2 & 1 \end{pmatrix}$$

The reader can verify that completing the square gives

$$\mathbf{T} = \begin{pmatrix} 1 & 2 \\ 0 & 1 \end{pmatrix} \quad \text{and} \quad \mathbf{D} = \begin{pmatrix} 1 & 0 \\ 0 & -3 \end{pmatrix}$$

or

$$Q(\mathbf{x}) = x_1^2 + 4x_1 x_2 + x_2^2 = (x_1 + 2x_2)^2 - 3x_2^2$$

which is positive when $|x_1 + 2x_2| > |x_2\sqrt{3}|$, negative when $|x_1 + 2x_2| < |x_2\sqrt{3}|$, and zero when $|x_1 + 2x_2| = |x_2\sqrt{3}|$. Section 5.14 shows that indefiniteness of y_{xx} can cause trouble for the N-R method, while negative definiteness will confound it totally when a minimum is sought.

Diagonalization of a singular matrix gives a quadratic form in a number of variables equal to the rank, necessarily less than N, the original number of variables. An example of this is $\mathbf{Q} = \begin{pmatrix} 1 & 1 \\ 1 & 1 \end{pmatrix}$, from which $Q(\mathbf{x}) = x_1^2 + 2x_1 x_2 + x_2^2 = (x_1 + x_2)^2$, and so $\mathbf{T} = \begin{pmatrix} 1 & 1 \\ 0 & 0 \end{pmatrix}$; $\mathbf{D} = \begin{pmatrix} 1 & 0 \\ 0 & 0 \end{pmatrix}$. Thus $Q(x)$ is positive semidefinite, vanishing when $x_1 = -x_2$ and being positive otherwise.

This section has defined the principal concepts regarding quadratic forms, giving examples in only two variables. The next shows a systematic way of performing the necessary computations when the number of variables is realistically large.

5.13. MULTIVARIABLE DIAGONALIZATION

Diagonalization of a quadratic form by completing the square becomes complicated when there are many variables. A more systematic way of performing the same transformation by the well-known Gauss elimination procedure for solving linear equations will now be described. This method takes less effort than solving a single set of N linear equations in N

unknowns, which compares quite favorably with the more widely known diagonalization technique of finding the eigenvectors of $\mathbf{Q}$ by factoring an Nth degree polynomial and then solving N sets of N linear equations. What makes Gauss elimination even more attractive is that its operations must be performed anyway during the routine generation of a N-R step. Thus the designer can determine the character of the hessian y_{xx} merely by paying attention to the numbers produced while solving $y_{xx}\Delta x = -y_x^T$.

Both methods—completing the square and Gauss elimination—will be demonstrated on a three variable example. Direct comparison will show that for each operation in one procedure there is a corresponding one in the other, proving that the two methods are equivalent. Since generalization to any number of variables is straightforward, the inductive formal proof of equivalence is omitted.

Consider then the three variable quadratic form

$$Q = x_1^2 + 2x_1x_2 + 4x_1x_3 + 3x_2^2 + 2x_2x_3 + 5x_3^2$$

Begin the process of completing the square by expressing as a perfect square all terms containing x_1 as a factor.

$$Q = (x_1 + x_2 + 2x_3)^2 + 2x_2^2 - 2x_2x_3 + x_3^2$$

Next write as a perfect square all terms in the remainder containing x_2 as a factor.

$$Q = (x_1 + x_2 + 2x_3)^2 + 2\left(x_2 - \tfrac{1}{2}x_3\right)^2 + \tfrac{1}{2}x_3^2$$

For a quadratic form in N variables, completing the square involves repetition of this process for $x_1,...,x_N$. Since in the example $N=3$, the procedure is finished. The form of the last equation suggests the following changes of variable:

$$\tilde{x}_1 \equiv x_1 + x_2 + 2x_3 \qquad \tilde{x}_2 \equiv x_2 - \tfrac{1}{2}x_3 \qquad \tilde{x}_3 \equiv x_3$$

so that $Q(\tilde{\mathbf{x}}) = \tilde{x}_1^2 + 2\tilde{x}_2^2 + \tfrac{1}{2}\tilde{x}_3^2$. Diagonalized in this way, the quadratic form is seen to be positive definite. The transformation and diagonal matrices $\mathbf{T}$ and $\mathbf{D}$ are

$$\mathbf{T} = \begin{bmatrix} 1 & 1 & 2 \\ 0 & 1 & -\tfrac{1}{2} \\ 0 & 0 & 1 \end{bmatrix} \qquad \mathbf{D} = \begin{bmatrix} 1 & 0 & 0 \\ 0 & 2 & 0 \\ 0 & 0 & \tfrac{1}{2} \end{bmatrix}$$

so $Q(\mathbf{x}) = \mathbf{x}^T\mathbf{Q}\mathbf{x} = \mathbf{x}^T\mathbf{T}^T\mathbf{D}\mathbf{T}\mathbf{x} = \tilde{\mathbf{x}}^T\mathbf{D}\tilde{\mathbf{x}} = Q(\tilde{\mathbf{x}})$, whence $\mathbf{Q} = \mathbf{T}^T\mathbf{D}\mathbf{T}$.

The procedure terminates as soon as there is no longer a remainder, which may occur before step N. This happens if and only if $\mathbf{Q}$ is singular, its rank being the number of variables already transformed when the remainder vanishes. If the coefficient of x_3^2 had been 4.5 instead of 5 in the example, there would have been no remainder after the second step, indicating that the form had rank two instead of three.

The procedure need not follow the order in which the variables are numbered, but a different order will produce a different transformation. Still the numbers respectively of positive, negative, and zero elements on the diagonal of $\mathbf{D}$ will be the same, since the quadratic forms must be equal before and after transformation. This pattern of signs, called the *signature* of Q, will always correctly identify the character of Q.

The variable in the remainder upon which the square is being completed must of course appear to the second power in one term. If there is no such term, as for x_1 in the quadratic form $x_1x_2 + x_2^2$, then some other variable with a second degree term (x_2 in the example) must be used, regardless of the order in which the variables are labeled. Whenever a remainder has no variable appearing to second degree, the remainder, and hence the entire quadratic form, is indefinite with respect to those variables. In an optimization problem any sign of indefiniteness—either mixed signs on the perfect squares or a remainder without second-degree terms—would require abandonment of the N-R method, making further completion of the square unnecessary.

Here's how completing the square can be accomplished systematically in the format of the well-known Gauss elimination technique for solving linear equations. Consider the application of this method to the hypothetical set of equations represented by $\mathbf{Qx} = \mathbf{b}$, where in the example

$$\mathbf{Q} = \begin{bmatrix} 1 & 1 & 2 \\ 1 & 3 & 1 \\ 2 & 1 & 5 \end{bmatrix}$$

and $\mathbf{b}$ is an arbitrary vector of constants (which need not be known). To eliminate the first variable x_1 from the second and third equations, Gauss elimination would divide the elements of the first row by the element in the first row and first column, called the *first pivot*. Appropriate multiples of the result are then added to the other rows to produce zeroes in their first columns, in the example giving the matrix

$$\begin{bmatrix} 1 & 1 & 2 \\ 0 & 2 & -1 \\ 0 & -1 & 1 \end{bmatrix}$$

This process is repeated for the next nonzero diagonal element, the second in the example, which gives a new matrix

$$\begin{bmatrix} 1 & 1 & 2 \\ 0 & 1 & -\frac{1}{2} \\ 0 & 0 & \frac{1}{2} \end{bmatrix}$$

Another application gives

$$\begin{bmatrix} 1 & 1 & 2 \\ 0 & 1 & -\frac{1}{2} \\ 0 & 0 & 1 \end{bmatrix}$$

the pivots being 1, 2, and $\frac{1}{2}$. The reader can verify that the last matrix is the same as the transformation matrix T obtained by completing the square. Moreover, the pivots are the diagonal elements of D arranged in the same order. The right-hand side vector b was not needed. Gauss elimination of Q gives the same result as completing the square because each operation in one corresponds exactly to an operation in the other, which the reader can verify directly in the example.

In the matrices generated by this method, the diagonal elements are the coefficients of second-degree terms in a single variable. To complete the square in this way then, only diagonal elements can be used as pivots, although Gauss elimination could in general employ any nonzero element for pivoting. Thus for the form $x_1 x_2 + x_2^2$, for which

$$Q = \begin{bmatrix} 0 & \frac{1}{2} \\ \frac{1}{2} & 1 \end{bmatrix}$$

only the diagonal element 1 could be used for a pivot, giving the matrix

$$\begin{bmatrix} -\frac{1}{2} & 0 \\ \frac{1}{2} & 1 \end{bmatrix}$$

Whenever all the remaining diagonal elements are zero, the elimination can stop because this proves the form indefinite wrt all variables remaining.

Section 4.12 showed that the N-R method required solving the linear equations whose matrix is the hessian matrix of second derivatives y_{xx}, evaluated at the point of departure. Thus to prove that y_{xx} is positive definite, one need only keep track of the pivots during the Gauss elimination used to solve the equations. Their positivity confirms that y_{xx} is positive definite. The next section shows why this is important.

5.14. TAMING THE N-R METHOD

Definiteness of the hessian y_{xx}, or lack of it, is what governs the performance of the N-R method. To see this, recall that the change in the quadratic approximation $a(\Delta x)$ for the step Δx is $-\frac{1}{2}(y_x)_0(y_{xx})_0(y_x)_0$. If $(y_{xx})_0$ is positive definite, then this change is negative whenever $(y_x)_0 \neq 0^T$, auguring improvement in an objective to be minimized. In this case as when $(y_{xx})_0$ is positive *semi*definite, the stationary point of the approximation is below the starting point, making exploration in the N-R direction attractive. But in all other situations, the stationary point might be above instead of below, in which case it is not worth seeking. Therefore the sign of $a(\Delta x)$ should always be checked before making the next step. If $a(\Delta x) < 0$, the predicted stationary point is below the start, and one can proceed with caution. But if $a(\Delta x) \geq 0$, a second-order method like N-R is misleading and should be abandoned in favor of something simpler like the gradient method, where improvement is guaranteed.

In solving the equations for the N-R step, one can determine, without additional calculation, the character of $(y_{xx})_0$ merely by inspecting the pivot elements. If they are all positive (negative), then $(y_{xx})_0$ is positive (negative) definite; if they are mixed, $(y_{xx})_0$ is indefinite; if any vanish, $(y_{xx})_0$ is singular. An indefinite matrix indicates that several stationary points may occur, warning that more than one direction of improvement needs to be examined.

Even when the approximation predicts improvement, the objective function at the new point should be computed to see if $y(x)$ really is better there. If it isn't, the approximation has deviated too far from y because of effects higher than second order, making the step overshoot the stationary point. To cope with this situation, shorten the step, say by making $(x)_1 = (x)_0 + \alpha \Delta x$ with $0 < \alpha < 1$. Improvement is guaranteed, because the step can be made small enough to make the approximation valid.

Armed with the theory presented so far, the designer can use the N-R method confidently. The semilog derivatives needed are easy to compute for the power functions encountered in most design problems. Yet the N-R method should be used only when the bounding techniques developed earlier cannot be applied, for when no lower bound is available to end computing, excessive iteration usually results.

5.15.   CONCLUDING SUMMARY

This chapter shows how to deal with minus signs in the objective. Negative terms cannot be condensed, but this doesn't prevent the condensation of any positive terms. Thus one can usually construct a bounding function that, having zero degrees of difficulty, can have no more than one easily found stationary point. When as in earlier chapters the lbf is a posynomial, this stationary point would be a minimum and consequently a valid lower bound for the objective. But this may not be true when there are negative terms, so second-order analysis is needed to see if the stationary point is a minimum. If it is, then the value of the lbf there bounds the objective below.

Checking the curvature at a stationary point requires completing the square on the second-order part of the Taylor expansion. The matrix of second derivatives needed is easy to compute in semilogarithmic form when the terms are power functions, that is, when the bounding function is a signomial. Consequently, global bounds for functions with several stationary points can be constructed.

When condensation cannot be used, the partial invariance method of the preceding chapters is still applicable. If degrees of difficulty obscure the situation, the easily computed first and second semilog derivatives can be employed for direct search by the N-R method. This can be prevented from converging to the wrong kind of stationary point by keeping an eye on the hessian matrix of second derivatives. The Gauss elimination scheme for solving the N-R equations will also show if the hessian is positive definite as required for minimization. Failure of this test signals that one is too far from the optimum to use second-order methods. Even then, first semilog derivatives are available for a gradient move. The N-R method can also be used on posynomials, which may be necessary when there are too many degrees of difficulty for condensation or partial invariance.

Thus a few negative signs in the objective may not be too troublesome, even though dual geometric programming cannot be used. In the constraints, however, negative terms can be dangerous, as the next chapter demonstrates.

NOTES AND REFERENCES

Signomials used to be called "generalized polynomials." Passy developed a stationary point theory for them and noted that dual geometric programming methods cannot be applied to them directly. Blau gave a N-R procedure for finding stationary points. Avriel and Williams devised a condensation scheme called "Complementary Geometric Programming" for computing stationary points which Avriel expanded into an elegant general theory for nonlinear programming (*op. cit.* Chapter 1). The name "signomial" was coined by Duffin and Peterson.

All of these are stationary point methods that can miss the global optimum, and none of them generate bounds for applying the principle of *satis quod sufficit*. Mancini combined the N-R method with interval arithmetic to identify regions in which a given stationary point is unique for a signomial. The global optimization method of this chapter, as well as the example, is in Barry McNeill's dissertation, which also contains other ideas on global optimization not yet published at this writing.

The Gauss elimination diagonalization procedure is adapted from *Foundations of Optimization (op. cit.* Chapter 1).

Avriel, M., and A. C. Williams, "Complementary Geometric Programming," *SIAM J. Appl. Math.*, **19**, 125–141 (1970).

Blau, G. E., and D. J. Wilde, "A Lagrangian Algorithm for Equality Constrained Generalized Polynomial Programming," *AIChE J.*, **17**, 235–240 (1971).

Duffin, R. E., and E. L. Peterson, "Geometric Programs with Signomials," Carnegie-Mellon Univ. Dept. Math. Report 70-38.

Mancini, L. J., and D. J. Wilde, "Kuhn-Tucker Regions in Signomial Programming," *Opns. Res.* **23**, sup. 1, Spr. 1975.

McNeill, B. W. "Tests for Global Optimality," unpublished doctoral dissertation, Stanford University, Stanford, Calif., 1976.

Passy, U., and D. J. Wilde, "Generalized Polynomial Optimization," *SIAM J. Appl. Math.*, **15**, 1344–1356 (1967).

EXERCISES

5-1.   Apply one step of the gradient method (Sec. 5.3) to the fertilizer plant of Sec. 5.1, using the base case of Sec. 5.4 as a starting point.

*5-2.   If your design project has a signomial objective and posynomial constraints, attempt to construct a bounding function having zero degrees of difficulty (Sec. 5.5).

5-3.   Verify that the semilog hessian matrix for the gravel box problem (Exercise 3-2) is positive definite (Secs. 5.6 and 5.13).

5-4.   Increase the gross income term for the fertilizer plant by a factor of 10 and solve the new problem.

5-5.   Apply one step of the Newton-Raphson method (Secs. 5.10, 5.11, and 5.14) to the fertilizer plant of Sec. 5.1, using the base case of Sec. 5.4 as a starting point.

5-6.   Minimize $y = 2x_1^2 x_2 - x_1^3 x_2 - x_1^2 x_2 - x_1 x_2 + x_2 + x_1^{-1} x_2 + x_2^{-1}$, starting at the base case $(1,1)^T$.

*Project problem.

5-7. (U. Passy) A chemical plant has reactor cost $3.18(10^{-4})x_1^{1.1}x_2^{0.6}$, separator cost $114x_2^{-1}x_3^{-1}$, catalyst cost $2.28x_3$, and income from byproduct sales of $3.38x_1^{0.25}$, where x_1 is reactor temperature, x_2 is reactor pressure, and x_3 is the weight fraction of catalyst used. Find the design giving the maximum profit.

5-8. Express W. Braga's van problem (Exercises 2-8 and 3-18) as one with possibly reversed posynomial constraints. Do not neglect anything, neither the resistance of the air nor the material cost. Find a satisfactory design.

CHAPTER 6

Conservation
and Global Optimization

*An optimist sees an opportunity in every calam-
ity;
a pessimist, a calamity in every opportunity.*

Anonymous

Past chapters have shown how posynomials, and some signomials, can
be bounded below by geometric mean inequalities. For this procedure to
be valid, any constraint functions must be bounded from above. Yet
important engineering problems have "reversed" constraints that, being
lower bounds, cannot be solved by the methods developed so far. Equivalently, a constraint bounded from above may have negative terms in it.
This will be seen to generate many stationary points and local minima,
forcing the designer to be very skeptical about accepting designs alleged to
be optimal. Such problems are not just mathematical oddities; they can
happen any time temperature is involved.

The inability of preceding methods to handle these difficulties is, at first
glance, a "calamity" as in the quotation. But this calamity is not without
opportunity for optimistic designers. This chapter will show that problems
with reversed inequalities may harbor unexpected local minima that may
turn out to be globally optimal. And lower-bounding functions, although
more difficult to construct than before, can still be obtained to confirm
global optimization.

Reversed constraints often arise from physical conservation laws for
mass or energy, so they are called here "conservation constraints." They

also occur when a shift of origin is necessary. One approach developed here is to use such constraints to eliminate variables from the objective. The nonposynomial form generated is then analyzed directly, there being but two major cases according to the value of a certain exponent. In one case local optima can occur at extreme values, and bounding methods are derived for detecting the global optimum from among several local ones. Engineering examples involve an irrigation reservoir, a cooled chemical reactor, and an ammonia storage system.

As in the unconstrained case of the preceding chapter, orthogonality conditions must be replaced with stationarity conditions, extended to the constrained case. Thus a limited amount of dual analysis is possible for identifying active constraints. Lower bounds can be constructed, but second-order analysis is required. And the N-R method can be extended to the constrained case when there are too many degrees of difficulty for the simpler methods.

The theory of this chapter makes optimization available for such design specialties as process design or thermodesign, where it has in the past been largely unapplied and occasionally misapplied. But pessimistic readers will recognize the calamity in these new found opportunities—they do require more care and computation.

6.1. CONSERVATION CONSTRAINTS

Many physical laws governing engineering design involve conservation of some quantity such as matter, energy, momentum, or distance. These conservation laws lead to linear equality constraints, usually with positive coefficients, of which the simplest example in two variables is $x_1 + x_2 = 1$. Despite its simplicity, such a constraint can cause difficulties whenever the objective function to be minimized is a nonlinear increasing function of one of the variables, say x_2, for then to bound the objective the constraint must be written as a posynomial bounded below by unity, with equality at the optimum.

$$x_1 + x_2 \geqq 1$$

Such an inequality is said to be *reversed* because it is in the wrong direction for condensation or geometric programming. The simplest form of such an objective is the sum of a posynomial in x_1 and a power function of x_2 having a positive exponent

$$y(x_1, x_2) = p(x_1) + x_2^{\theta}$$

with $\theta > 0$ and $0 < x_1 \leq 1$.

One recommended approach to this problem is to use the conservation constraint to eliminate the variable x_2 in the power function. Being linear, the conservation constraint is easily solved for x_2: $x_2 = 1 - x_1 \geq 0$. Elimination of x_2 from y gives

$$y(x_1) = p(x_1) + (1 - x_1)^\theta$$

The first derivative is

$$\frac{\partial y}{\partial x_1} = \frac{\partial p}{\partial x_1} - \theta (1 - x_1)^{\theta - 1}$$

At $x_1 = 1$, $\partial p / \partial x_1$ is finite, but the behavior of $-\theta (1 - x_1)^{\theta - 1}$ depends on θ. If $\theta \geq 1$, the derivative of $(1 - x_1)^\theta$ is finitely negative, but when $0 < \theta < 1$, the derivative is *in*finitely negative. Hence there is *always* a local minimum at $x_1 = 1$ when $\theta < 1$.

The second derivative is

$$\frac{\partial^2 y}{\partial x_1^2} = \frac{\partial^2 p}{\partial x_1^2} + \theta (\theta - 1)(1 - x_1)^{\theta - 2}$$

If the second derivative is positive, as is the second derivative of $(1 - x_1)^\theta$ when $\theta \geq 1$, then y has only one local minimum, which therefore must be a global minimum. This point may be at one of the simple bounds, either $x_1 = 0$ or $x_1 = 1$, or it may be in the interior, but there will be no other unconstrained stationary points (see Fig. 6-1). Section 6.4 shows how to construct lower bounds for such functions.

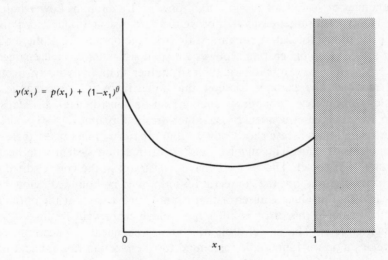

Figure 6-1 Conservation: $\theta \geqslant 1$

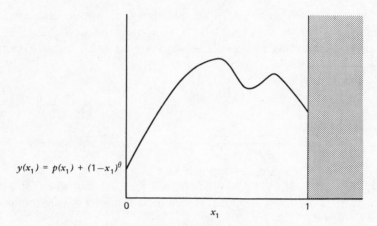

$$y(x_1) = p(x_1) + (1-x_1)^\theta$$

Figure 6-2   Conservation: $\theta \leqslant 1$

If $\theta < 1$, the second derivative of y does not have a fixed sign. This situation, which already leads to a constrained minimum at $x_1 = 1$, may have interior minima as well. It can also have interior maxima, and in some circumstances it will even have a constrained minimum at $x_1 = 0$ (see Fig. 6-2). Bounding and minimizing such a function takes special care, but the next section will show how to do it.

The next three sections apply this theory to two engineering design problems. The first involves a pipeline and reservoir system in which economies of scale lead to more than one local minimum. Lower bounds are constructed rather easily that prove the base case at a boundary to be a global, not just a local, optimum. Aside from the optimization, the modeling of the reservoir cost as a power function will interest civil engineers, who presently size reservoirs graphically rather than analytically. Section 6.3 modifies this example to show that when there are *dis*economies of scale, a linear lower-bounding function can be constructed. This leads to another example of condensation, rather drastic this time.

A second design case shows how to find interior minima when there are economies of scale. This involves a chemical reaction system with heater, reactor, and cooler. This problem has a constraint in the correct direction for condensation, but the conservation term must be bounded below by a linear secant producing an error that cannot vanish, even at the optimum. In the example this error is not large enough to prevent finding a good design anyway. The model itself will interest chemical engineers, since it develops a power function lower bound for the reaction rate integral used in reactor design.

6.2. MODELING AN IRRIGATION RESERVOIR

This section is a sequel to the pipeline design problem of Secs. 3.14–3.16. Imagine now that the pipeline is to supply irrigation water to a discharge reservoir intended to contain the fluctuations in seasonal demand. The design evaluated in Chapter 3 was for the flow fixed at the annual average. One can, however, design the line for a larger flow and shut it down occasionally when the reservoir is full. This would cost more for the pipeline, but less for the smaller reservoir. The model is developed and discussed in this section; the next works out the solution and general conclusions.

To keep things parametric at first, let the average annual demand flow rate be A; the peak demand, P. Then the design flow rate q to be chosen is bounded by A and P: $A \le q \le P$. Modeling the reservoir cost as an algebraic function of the flow is interesting because presently the design method is entirely graphical. The designer must have a plot of cumulative water volume to be delivered as a function of elapsed time during the year (Fig. 6-3). The slope of this S-shaped curve is steepest during the dry

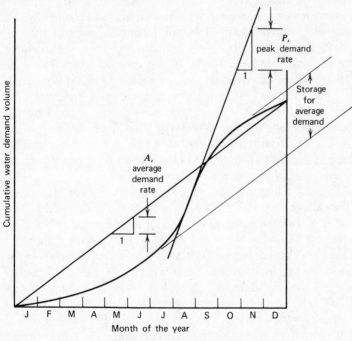

Figure 6-3 Reservoir sizing

summer when water is most needed; this slope is P, the peak demand rate. The slope of the straight line between one end of the curve and the other is the average annual demand A, the least for which the pipeline can be designed, which here corresponds to the base case. The volume of the reservoir needed for a pipeline pumping at this average rate is determined graphically by drawing lines parallel to the average demand line just touching the demand curve, both above and below the average line. The vertical distance between these parallels is the reservoir volume required to contain the difference between actual and average demand. This present graphical design method determines the base case. The optimization problem is to see if this design is indeed globally optimal, as implicitly assumed by the present method.

For an optimization model, the reservoir cost must be expressed as a function of the flow, so flows other than the annual average have to be considered. The peak demand rate P is simply the steepest slope of the demand curve, and the reservoir cost at that flow is zero. In practice of course, there must be some small reservoir even in this case, but the designer is concerned only with the variable cost above this minimum, so the variable cost is zero when the pipeline can meet the maximum demand rate. Thus a function of the form $R(P\text{-}q)^\theta$ was proposed for the reservoir cost, R and θ being parameters to be determined empirically, the peak demand being known already. Two points, one for $q = A$ and the other for $q = P$, have already had their costs evaluated; at least one more is needed to obtain R and θ independently. In the case on which this example is based, the designer determined reservoir volumes for not just one, but two intermediate rates and estimated the corresponding reservoir costs from maps. Since $\log R(P - q)^\theta = \log R + \theta \log(P - q)$, the logarithm of the reservoir cost was plotted against $\log(P - q)$ as in Fig. 6-4. The straightness of the resulting line confirmed the model, which was remarkable considering that the cost depends on such natural nonlinear (and non power function) phenomena as rainfall and land topography.

This cost not being in standard power function form, it is tempting to make the change of variable $u \equiv P - q$ so that the cost would be the power function Ru^θ. But since the cost is being minimized, the transformation equation would have to be written $P^{-1}u + P^{-1}q \geq 1$ for it to be active at the minimum. This conservation of mass constraint is a reversed inequality, and so the original unconstrained form $R(P - q)^\theta$ is retained to make the theory of the preceding section applicable. In the example θ was found to be $0.90 < 1$, so it is immediately established that there is a local minimum at $q = P$, which is *not* the base case. Physically speaking, this says that very small reservoirs are less economical than no reservoir at all. It warns the designer that the unconventional design having a large

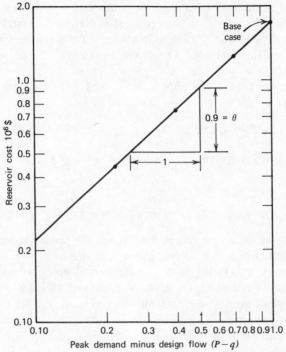

Figure 6-4   Reservoir cost

pipeline and no discharge reservoir cannot be discredited out of hand, even though such a design would boggle the mind of any experienced irrigation engineer.

Besides the reservoir cost, there are several other expenses that vary with flow, although piping and station costs do not. The cost of overcoming friction now has two contributions. First is the capital cost for that part of the pump capacity needed for overcoming friction head, as opposed to hydrostatic head. This is the amortized *friction capital cost* $F_c d^{-\alpha} q^{\eta}$, where F_c is a known parametric constant, α is a known exponent derived from the fluid mechanics, as in Chapter 3, and the known exponent η also depends on the fluid (it is 2.85 for water). Second is the cost of the energy lost by friction. This is proportional to the friction capital cost, except that it must be multiplied by A/q, the fraction of the time that the line is in operation. Let all constant coefficients be combined into a single parameter F_e so that the *friction energy cost* is $F_e d^{-\alpha} q^{\eta-1}$.

A final cost is that of pumping the fluid against the static head. The energy cost for doing this is constant, independent of the design variables, since a fixed amount of water must be pumped in a year. Therefore it need not be included in the variable cost. However, the capital cost of the pumps does depend on the design flow rate, since the pumps must be paid for even when they are idle. This *static capital cost* is Hq, where H is a known parameter.

The problem then is to choose n, d, and q minimizing the annual variable cost

$$c(n,d,q) \equiv Sn + Bd^{2\beta} + Ed^{2\beta}n^{-\gamma} + F_c d^{-\alpha}q^{\eta} + F_e d^{-\alpha}q^{\eta-1} + Hq + R(P-q)^{\theta}$$

The flow rate is constrained to the interval $A \le q \le P$.

6.3. GLOBAL MINIMIZATION WITH A CONSERVATION CONSTRAINT

In the pipeline-reservoir system modeled in the preceding section, let the flow be scaled so that the annual average is at $q(\equiv x_3) = 1$, and the peak demand is at $x_3 = 2.00$. As in Sec. 3.16, partial optimization when $q = 1$ gives a pipe diameter of 63, the maximum allowed, and the optimal integer number of stations is five. To scale the problem to this base case, let $x_1 \equiv d/63$; $x_2 \equiv n/5$; $y \equiv c/5.74(10^6)$. With numerical values introduced for the parameters, the scaled variable cost is

$$y = 0.131x_2 + 0.264x_1^{1.80} + 0.158x_1^{1.80}x_2^{-1} + 0.028x_1^{-4.87}x_3^{2.85}$$
$$\text{stations} \quad \text{std. pipe} \qquad \text{extra pipe} \qquad \text{friction capital}$$

$$+ 0.141x_1^{-4.87}x_3^{1.85} + \quad 0.1044x_3 + 0.174(2.00 - x_3)^{0.90}$$
$$\text{friction energy} \quad \text{static cptl.} \qquad \text{reservoir}$$

Since the exponent in the reservoir cost is less than unity, the objective may have several local minima. The constraints are simple bounds: $x_1 \le 1$ and $1 \le x_3 \le 2$. The designer wishes to see if this base case gives the globally optimal cost.

To find out, bound the reservoir cost as follows:

$$0.174(2 - x_3)^{0.9} = 0.174(2^{0.9})\left(1 - \tfrac{1}{2}x_3\right)^{0.9}$$
$$\ge 0.325\left(1 - \tfrac{1}{2}x_3\right) = 0.325 - 0.162x_3$$

There is already a linear term (static capital) in the objective, which can be combined with that in the lower bound to give $0.1044x_3 + 0.174(2 - x_3)^{0.9} \ge 0.325 - 0.058x_3$. The linear term is still negative, and a posynomial lower bound is desired. Hence the term with the smallest exponent on x_3 is compared with the right member of the last inequality

$$0.141x_1^{-4.87}x_3^{1.85} + 0.1044x_3 + 0.174(2 - x_3)^{0.9}$$
$$\geq 0.141x_1^{-4.87}x_3^{1.85} + 0.325 - 0.058x_3$$

But since $x_1 \leq 1$, it follows that $x_1^{-4.87} \geq 1$. Moreover, $x_3 \geq 1$ implies $x_3^{1.85} \geq x_3$, and so $0.141x_1^{-4.87}x_3^{1.85} \geq 0.141x_3$. Combining these inequalities gives

$$0.141x_1^{-4.87}x_3^{1.85} + 0.1044x_3 + 0.174(2 - x_3)^{0.9}$$
$$\geq 0.141x_3 + 0.325 - 0.058x_3$$
$$= 0.325 + 0.083x_3$$

Therefore

$$y \geq 0.131x_2 + 0.264x_1^{1.80} + 0.158x_1^{1.80}x_2^{-1} + 0.028x_1^{-4.87}x_3^{2.85} + 0.083x_3 + 0.325$$

with equality at the base case where $\mathbf{x} = \mathbf{1}$.

This lbf is a posynomial as desired. But even more important, it strictly increases with x_3. Therefore its global minimum is the constrained local minimum where x_3 is as small as possible—at the base case. Hence the original objective must have its global minimum there also, even though the point where $x_3 = 2$ is known to be a local minimum. This bounding procedure has identified the global minimum of a multimodal function without searching or even taking a derivative!

By itself, the monotonicity of the lower bounding function wrt x_3 is not enough to prove global minimality, since there are two more variables x_1 and x_2 involved. But the base case was constructed in Sec. 3.15 by partially optimizing wrt x_1 and x_2 while x_3 was fixed at its base value. Since this partial optimization problem is a posynomial geometric program, the result is a global minimum wrt x_1 and x_2 for the base value of x_3, now known to be optimal. Therefore the base case is globally minimal wrt all three variables.

A rough design generalization can now be made. In irrigation systems, where the cost of storage is relatively low and large volumes must be moved, the globally optimal flow tends to be the smallest feasible, namely the annual average demand. On the other hand, a petroleum pipeline may be globally optimal when designed for peak flow, for then the expensive steel tankage is as small as possible. It behooves the designer to choose the base case accordingly and check it for global optimality by the bounding method described.

Computer codes for geometric programming being now widely available, it would have been tempting to write the problem as a posynomial with a reversed inequality. Codes based on the N-R method would then converge to an interior stationary point, which in this example is the global *maximum* rather than the minimum. Beware of reversed inequalities; they can turn optima into pessima.

The case where the global minimum is interior instead of at a bound is discussed in Sec. 6.5 in a chemical engineering context. It is convenient first to study the preceding problem when the reservoir exponent exceeds unity.

6.4. DISECONOMY OF SCALE

The situation just discussed, in which the exponent of the conservation term is less than unity, would be characterized by economists as having "economies of scale." This is because the cost increase is less than proportional to the independent variable, $[1-(q/2)]$ in the example. Strict proportionality (called the *linear case*) would have $\theta = 1$, and $\theta > 1$ is said to exhibit "diseconomies of scale." Thus the linear case, which is a secant lbf when $\theta < 1$, as in the preceding example, becomes an ubf when $\theta > 1$ and therefore cannot be used to find a satisfactory design. Figure 6-5 illustrates this.

However, a different linear lbf can be constructed when $\theta > 1$, because then the function lies entirely above any tangent, except of course at the point of tangency. To see this, let $f(x) \equiv (K-x)^{\theta}$, with $\theta > 1$ and $K > 1$. The constant K has been introduced to handle the situation where, as in the example, the problem is scaled so that $x = 1$ in the base case. Section 6.1 proved that the second derivative of $f(x)$ is always positive for positive x and $\theta > 1$. Thus the Taylor expansion to second order gives, for any value x_0 in the interval $1 \le x_0 \le K$,

$$f \approx f_0 + \left(\frac{\partial f}{\partial x}\right)_0 (x-x_0) + \tfrac{1}{2}\left(\frac{\partial^2 f}{\partial x^2}\right)_0 (x-x_0)^2 \ge f_0 + \left(\frac{\partial f}{\partial x}\right)_0 (x-x_0)$$

with equality if and only if $x = x_0$. This can be interpreted geometrically as $f(x)$ being above any tangent (see Fig. 6-5). Mathematicians would speak of $f(x)$ as being *strictly convex* in this situation, which always occurs when the second derivative is positive in the region of interest.

The right member of the inequality is therefore a lbf for the conservation term. It is linear, and its coefficients are readily evaluated from the first

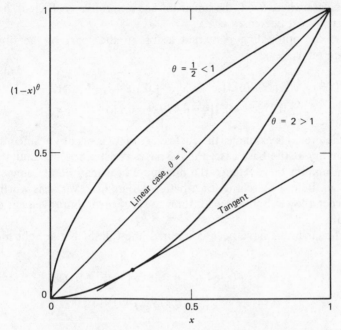

Figure 6-5 Secant and tangent lower-bounding functions

derivative at x_0

$$\left(\frac{\partial f}{\partial x}\right)_0 = -\theta(K-x_0)^{\theta-1} = -\theta f_0/(K-x_0)$$

Therefore

$$f \geq f_0\left(1 + \frac{\theta x_0}{K-x_0}\right) - \left(\frac{\theta f_0}{K-x_0}\right)x$$

with equality if and only if $x = x_0$.

As an illustration, modify the irrigation example so that the reservoir cost coefficient is increased 150% and the exponent is made 1.10 instead of 0.90. Then

$$0.435(2-x_3)^{1.10} \geq 0.435\left(1 + \frac{1.10(1)}{(2-1)}\right) - \frac{1.10(0.435)}{(2-1)}x_3$$

$$= 0.914 - 0.479x_3$$

As in the case where $\theta < 1$, the linear lbf has a negative coefficient, because the reservoir cost decreases wrt x_3.

To test for optimality, construct a lbf on that part of the objective depending on x_3.

$$0.028x_1^{-4.87}x_3^{2.85} + 0.141x_1^{-4.87}x_3^{1.85} + 0.1044x_3 + 0.435(2 - x_3)^{1.10}$$
$$\geq 0.273x_3 + 0.914 - 0.479x_3 = 0.914 - 0.206x_3$$

Unlike the previous example, the coefficient of x_3 is negative. Since there is strict equality at the base case, the lbf has a constrained maximum rather than a minimum there. Hence the minimum cost may be at some greater value of x_3. Before searching for a better design, however, it is worthwhile to construct a lower bounding constant to see if much improvement can be expected.

After linear terms have been combined, the full lbf is the right member of

$$y \geq l(\mathbf{x}) \equiv \left(0.131x_2 + 0.264x_1^{1.80} + 0.158x_1^{1.80}x_2^{-1} + 0.028x_1^{-4.87}x_3^{2.85}\right.$$
$$\left. + 0.141x_1^{-4.87}x_3^{1.85} - 0.375x_3\right) + 0.914$$

There is also a diameter constraint that must be tested to see if it remains tight, as it is in the base case. The lower bound on the flow x_3, tight in the base case, is now known to be loose at the minimum, so it will not be written. The upper bound on x_3 is not written either, since it can easily be checked later. There being in all seven terms for three variables, a lower bound can be constructed by condensing the posynomial part of the objective to two terms. This must be done carefully to avoid excessive distortion leading to unboundedness.

First consider the first and third terms, the only ones containing x_2. Partial optimization wrt x_2 would produce a zero degree of difficulty problem making the two terms equal. In fact, they are unequal in the base case only because of the integer constraints on pipe diameter and number of stations. Hence let them be condensed into one term using equal weights rather than those of the base case.

$$0.131x_2 + 0.158x_1^{1.80}x_2^{-1} \geq \left(\frac{0.131x_2}{\frac{1}{2}}\right)^{1/2}\left[\frac{0.158x_1^{1.80}x_2^{-1}}{\frac{1}{2}}\right]^{1/2}$$
$$= 0.287x_1^{0.90}$$

Notice that x_2 has cancelled out because of this choice of weights. Since the new condensed posynomial has one less term and one less variable, its number of degrees of difficulty is unchanged. Hence it must be condensed all the way down to a single term. The base case weights will be used.

$$0.287x_1^{0.90} + 0.264x_1^{1.80} + 0.028x_1^{-4.87}x_3^{2.85} + 0.141x_1^{-4.87}x_3^{1.85}$$
$$\geq 0.720x_1^{-0.124}x_3^{0.473}$$

The lbf resulting is the right member of the following inequality, in which λ is an undetermined nonnegative exponent taking account of the constraint $x_1 \leq 1$. $(1 \geq x_1^\lambda)$

$$y \geq 0.914 + \left(0.720x_1^{-0.124}x_3^{0.473} - 0.375x_3\right)x_1^\lambda$$

At the unique stationary point of this lbf, the terms must satisfy the following stationarity conditions

$$
\begin{aligned}
x_1: &-0.124t_1 &+\lambda(t_1 - t_2) &= 0 \\
x_3: &\ \ 0.473t_1 - t_2 & &= 0
\end{aligned}
$$

The first of these shows that λ must be strictly positive, which implies that the constraint $x_1 \leq 1$ must be tight at the stationary point. Hence x_1 must be set to unity and the objective recondensed.

With $x_1 = 1$, the bounding inequality posynomial becomes

$$0.551 + 0.028x_3^{2.85} + 0.141x_3^{1.85}$$
$$\geq 0.551 + 0.169x_3^{2.016}$$

The new lbf is the right member of the following inequality

$$y \geq 1.465 + \left(0.169x_3^{2.016} - 0.375x_3\right)$$

The zero-degree-of-difficulty signomial in parentheses has a unique stationary point where $2.016t_1 - t_2 = 0$, from which

$$2.016t_1 = 2.016\left(0.169x_3^{2.016}\right) = 0.375x_3 = t_2$$

Therefore

$$x_3 = \left(\frac{0.375}{2.016(0.169)}\right)^{1/1.016} = 1.099 < 2 = K$$

Since this is less than the upper bound on x_3, this stationary point is inside the feasible interval. The second semilog derivative of the lbf is $(2.016)^2 t_1 - t_2 = 4.06(0.204) - 0.412 > 0$, so the stationary point is a minimum. Evaluating the lbf there gives a lower bound on the cost

$$y \geq 1.257$$

The objective function was not scaled in this example, so its value must be determined in the base case

$$y(1) = 1.261$$

Thus the base case, even though not proven to be the minimum, is within less than 1% of the global minimum

$$1.261 \geq y_{\star} \geq 1.257 \qquad \textit{satis quod sufficit}$$

At x_1, where the lbf is minimum, the objective function is 1.268, somewhat worse than at the base case. If the gap between base case objective and lower bound had been large enough to require continuing the search, then the N-R procedure of Chapter 5 would have been needed to find a better design between $(x_3)_0 = 1$ and $(x_3)_1 = 1.099$.

6.5. ECONOMY OF SCALE IN A CHEMICAL REACTION SYSTEM

A new example will show how to handle economies of scale ($\theta < 1$) when the global minimum is interior rather than at a boundary. The annual variable cost $c(\xi)$ of a chemical reaction system, including heater, reactor, separator, and auxiliary cooler, is to be minimized wrt the extent of reaction ξ, which ranges from zero, when there is no reaction, to unity when the reaction is complete (see Fig. 6-6). The base case is a system

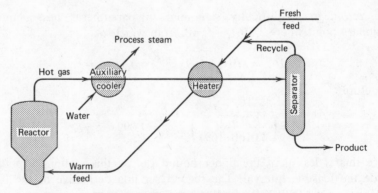

Figure 6-6 Chemical-reaction system

without an auxiliary cooler, for which the extent of reaction is $\xi_t = 0.254$, corresponding to thermal equilibrium in which the heat of reaction given off just balances the amount needed to bring the reactants up to the reaction temperature.

Feed heater capital and energy costs are, respectively, $15.0(10^3)\xi^{-0.60}$ and $6.55(10^3)\xi^{-1}$. The reactor cost is a power function of the reactor volume v, given by the integral equation

$$v = K\xi^{-1} \int_0^\xi r^{-1}(x)\,dx$$

where the integrand is the reciprocal of the positive difference between the forward and reverse reaction rates for the isothermal reaction: $r(\xi) = p_f(\xi) - p_r(\xi) > 0$. These rates are in turn ratios of polynomials in ξ, so the complete reactor cost expression is far from the power function form of the models developed so far. Yet in the case studied here, the reactor cost is shown in the *American Institute of Chemical Engineers Journal* to be bounded below by the following expression, a power function of $1-\xi$, the fraction not reacting: $89.0(10^3)(1-\xi)^{-0.313}$. To simplify the example, this lbf is used instead of the reactor cost, and in actuality the error turns out to be negligible at the optimum. This may be written $89.0(10^3)u^{-0.313}$ with $u \le 1-\xi$, which gives the standard posynomial constraint $\xi + u \le 1$.

An auxiliary cooler is required if the reaction rate exceeds that giving thermal equilibrium ($\xi > \xi_t = 0.254$). The annual cost then increases wrt the difference $\xi - \xi_t$, there being economies of scale in the expression $33.0(10^3)(\xi - \xi_t)^{0.60}$. There is also a credit for the steam generated in this waste heat boiler, appearing as a negative cost $-3.40(10^4)(\xi - \xi_t)$.

In the base case, with the scaled variables $x_1 \equiv \xi/\xi_t = \xi/0.254$, $x_2 \equiv u/(1-\xi_t) = u/0.746$, and $y \equiv c/157.5(10^3)$, the scaled objective and constraints are

$$y = 0.217x_1^{-0.60} + 0.164x_1^{-1} + 0.619x_2^{-0.313}$$
$$0.254x_1 + 0.746x_2 \le 1$$

Here the auxiliary cooler cost is of course zero, since none is required in the base case. In terms of the scaled variable x_1, this auxiliary cooler cost, with steam credit for the waste heat recovered, is, when divided by the reactor and heater base cost $157.5(10^3)$

$$0.210(\xi - \xi_t)^{0.60} - 0.0216(\xi - \xi_t)$$
$$= 0.210(0.254)^{0.60}(x_1 - 1)^{0.60} - 0.0216(0.254)(x_1 - 1)$$
$$= 0.0923(x_1 - 1)^{0.60} - 0.0549x_1 + 0.0549$$

The complete scaled objective function is

$$y = 0.217x_1^{-0.60} + 0.164x_1^{-1} + 0.619x_2^{-0.313}$$
$$+ 0.0923(x_1 - 1)^{0.60} - 0.055x_1 + 0.055$$

with the active constraint $0.254x_1 + 0.746x_2 \leqq 1$. Since the accuracy of the model is certainly no better than two significant figures, let the satisfaction threshold be $\tau = 0.01$.

Since the conservation term has a constant subtracted from the variable, just the opposite situation from that studied in this chapter's previous sections, some analysis is in order. First check to see if the substitution $w \equiv x_1 - 1$ gives a standard posynomial program. Since the cooler cost would increase wrt x_1, the substitution equation is replaceable by the active inequality $w \geq x_1 - 1$. This is transformable into the posynomial inequality $x_1^{-1} + wx_1^{-1} \geqq 1$, which is reversed and hence inappropriate for geometric programming. As in previous sections, it is recommended *not* to make this substitution. The cooler cost will be therefore left in its original form.

The first semilog derivative of the cooler cost, called t_4, is $\partial t_4 / \partial \ln x_1 = \theta t_4 x_1 (x_1 - 1)^{-1}$. This becomes infinite as x_1 approaches 1 from above, indicating that y must have a local minimum at the base case. In fact, the second derivative is negative for $x_1 > 1$

$$\frac{\partial^2 (x_1 - 1)}{\partial x_1^2} = \theta(\theta - 1)(x_1 - 1)^{\theta - 2} = \theta(\theta - 1)t_4(x_1 - 1)^{-2} < 0$$

for $0 < \theta < 1$. Hence the function may be bounded below by a secant connecting the ends of the range $1 \leq x_1 \leq 3.94[=(0.254)^{-1}]$, as shown in Fig. 6-7. The inequality is

$$(x_1 - 1)^{0.60} \geq \frac{1.910}{(3.94 - 1)}(x_1 - 1)$$

where $1.910 = (3.94 - 1)^{0.60}$

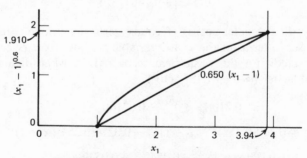

Figure 6-7 Secant lower-bounding function

The auxiliary cooler cost is therefore bounded below by a linear expression

$$0.0923(x_1 - 1)^{0.60} \geq 0.0600x_1 - 0.0600$$

With the steam credit included, the net charge to the auxiliary cooler is bounded below by

$$0.0923(x_1 - 1)^{0.6} - 0.0549x_1 + 0.0549 \geq 0.0051x_1 - 0.0051$$

Hence a lbf for y is the right member of

$$y \geq l_0 \equiv \left(0.217x_1^{-0.60} + 0.164x_1^{-1} + 0.619x_2^{-0.313} + 0.0051x_1\right) - 0.0051$$

with the constraint $0.254x_1 + 0.746x_2 \leqq 1$.

The variable part of l_0, together with the constraint, has three degrees of difficulty. Scaling the variable part gives the weights needed for a first condensation

$$l_0 = 1.005\left(0.216x_1^{-0.60} + 0.163x_1^{-1} + 0.616x_2^{-0.313} + 0.0051x_1\right) - 0.0051$$

If any three primal weights are used to estimate dual weights, the normality condition guarantees that the fourth dual weight will equal the fourth primal weight. Therefore

$$\delta_{01} = w_1 = 0.216 \qquad \delta_{02} = w_{02} = 0.163$$
$$\delta_{03} = w_{03} = 0.616 \qquad \delta_{04} = w_{04} = 0.0051$$

The orthogonality conditions are:

$$\begin{aligned} x_1: & -0.60\delta_{01} - \delta_{02} & + \delta_{04} + \delta_{11} & = 0 \\ x_2: & & -0.316\delta_{03} & + \delta_{12} = 0 \end{aligned}$$

from which

$$\delta_{11} = 0.60\delta_{01} + \delta_{02} - \delta_{04} = 0.60w_{01} + w_{02} - w_{04} = 0.288$$
$$\delta_{12} = 0.313\delta_{03} = 0.313w_{03} = 0.193$$

Then $\lambda_1 = \delta_{11} + \delta_{12} = 0.480$, and $w_{11} = \delta_{11}/\lambda_1 = 0.706$; $w_{12} = \delta_{12}/\lambda_1 = 0.294$. The dual lower bound on the variable part of the lbf is

$$\left(\frac{0.216}{0.216}\right)^{0.216}\left(\frac{0.163}{0.163}\right)^{0.163}\left(\frac{0.616}{0.616}\right)^{0.616}\left(\frac{0.0051}{0.0051}\right)^{0.0051}\left(\frac{0.254}{0.706}\right)^{0.288}\left(\frac{0.746}{0.294}\right)^{0.193}$$

$$= \left(\frac{0.254}{0.706}\right)^{0.288}\left(\frac{0.746}{0.294}\right)^{0.193} = 0.881$$

Therefore $l_0 \geq 1.005(0.881) - 0.0051 = 0.885 - 0.0051 = 0.880$. The gap between y_0 and this lower bound being $1 - 0.88 = 0.12 > 0.01$, the interior must be examined for possible minima.

A new design is generated from the constraint weights: $w_{11} = 0.706 = 0.254(x_1)_1$, whence $(x_1)_1 = 2.36$. Similarly, $(x_2)_1 = 0.294/0.746 = 0.54$. At the new design x_1, the objective is 0.987, more than 1% below the base case value. Therefore the base case is *not* the global minimum, although it is known to be a *local* minimum! Remarkably, this was deduced from information at the base case, together with one evaluation of y at x_1.

At x_1, the lbf $l(x_1)$ is 0.958, and a lower bound on this lbf, obtained as in the base case but with dual variables generated by the new design, is 0.942. Thus the lbf at x_1 is not sufficiently close to its minimum. However, further search for the minimum of l is not worth the trouble, since much of the gap between y and l is due to approximation error, which is 3% at x_1. It is better to bracket the region containing the local minimum and make a more accurate approximation in this smaller interval.

The next two sections show how to find a local minimum by the N-R method, which would usually be necessary in problems with several degrees of freedom. But in the present problem, having only one degree of freedom, it is simpler to bracket the local minimum with a few direct evaluations of the objective.

Some cases are now evaluated to locate a region containing the minimum. At a new point $x_2 \equiv (2.00, 0.660)$, the objective is $y_2 = 0.968 < 0.987 = y_1$, so x_1 is decreased again. At $x_3 \equiv (1.70, 0.762)$, $y_3 = 0.964 < y_2$, leading to another decrease in x_1. At $x_4 \equiv (1.40, 0.864)$, $y_4 = 0.974 > y_3$, so it is now known by continuity that there is at least one local minimum x_* in the interval

$$1.40 < (x_1)_* < 2.00$$

Before checking this smaller region, it is wise to examine the other two regions for possibly better designs. The discussion of how to do this will be sketchy here so that the reader can carry out the details as a set of instructive exercises in constructing secant bounds and computing bounding constants.

Consider first the left interval $1.00 \leq x_1 \leq 1.40$. The lbf constructed is, when scaled to $x_1 = 1.40$,

$$y \geq l_1 \equiv 1.052(0.169x_1^{-0.60} + 0.111x_1^{-1} + 0.619x_2^{-0.313} + 0.104x_1) - 0.078$$

A lower bound constructed from the weights at $x_1 = 1.40$ is 0.965, slightly greater than the objective value at $x_3 = (1.70, 0.762)$, namely 0.964. This proves there can be no global minimum in the left interval.

·

The right interval $2.00 \le x_1 \le 3.94$ is slightly harder to analyze because the lbf has a negative term.

$$y \le l_3 \equiv 0.217 x_1^{-0.60} + 0.164 x_1^{-1} + 0.619 x_2^{-0.313} - 0.0116 x_1 + 0.0606$$

To have zero degrees of difficulty, all positive terms in the objective must be condensed into a single term, as must the two constraint terms. The result, using weights at $x_2 = (2.00, 0.660)$, is

$$l_3 \ge l_3' \equiv 0.955 x_1^{-0.181} x_2^{-0.237}$$

and

$$1 \ge 0.254 x_1 + 0.746 x_2 \ge 0.863 x_1^{0.508} x_2^{0.492}$$

To the reader is left the exercise of proving that l_3' is minimum in this interval at x_2, which eliminates the right-hand region as a place for the global minimum. It is now known for sure that the *global* minimum x_* must be in the center interval:

$$1.40 < (x_1)_* < 2.00$$

The lbf l_2 for this center interval, scaled to the best design so far at $x_2 = (1.70, 0.762)$, is

$$y \ge l_2 \equiv 0.946 \left(0.167 x_1^{-0.60} + 0.102 x_1^{-1} + 0.713 x_2^{-0.313} + 0.018 x_1 \right) + 0.017$$

The approximation error at x_2 is only 0.002, which is acceptably small compared to the satisfaction threshold of 0.01. A lower bounding constant for l_2, constructed from the weights at x_2, is 0.929, much lower than $y_2 = 0.963$. Hence a better design or a better lower bound must be sought.

The reader can verify that the new design generated by the dual weights used in constructing the lower bound is $x_5 \equiv (1.72, 0.734)$. The objective there is $y_5 = 0.964 = y_2$, so the new design is just as good as the best already available. However, l_2 is smaller at x_5, being 0.956 rather than 0.963, so the weights at x_5 are used to construct a new lower bound on l_2. The reader can verify that this bound is 0.955, only 0.008 less than the best objective value. Thus the design $x_2 = (1.70, 0.762)$ falls within the *global* satisfaction threshold. *SQS*.

With slight modification, this approach can handle the more difficult and realistic case where a fixed construction charge must be considered. Thus in the example any value of x_1 other than that at the base case requires an auxiliary cooler which can have, in addition to the variable costs already considered, a fixed cost independent of size. This would be to cover things like a control system, safety devices, and supporting structure. Suppose this fixed cost to be $20,000, which is 13% of the variable cost at the base case. Then this constant 0.13 must be added to the objective function whenever $x_1 > 1$. The design x_2 obtained would then cost $0.96 + 0.13 = 1.09$, ruling it out as a design better than the base value. In fact, the first lower bound constructed would, with this fixed charge added, be $0.94 + 0.13 = 1.07 > 1$, which would immediately prove the base case to be globally optimal.

This problem was special in having only one degree of freedom, which made it easy to locate, or at least bracket, the local minimum with a few direct evaluations of the objective function. But this is not practical when there are several degrees of freedom, so the next two sections show how to extend the powerful N-R method to the general signomial optimization problem with constraints.

6.6. CONSTRAINED STATIONARITY CONDITIONS

Chapter 4 showed how useful are the orthogonality conditions in constrained posynomial problems. Then Chapter 5 gave stationarity conditions, based on setting logarithmic derivatives to zero, analogous to the orthogonality conditions and usable for signomial objectives. The present section develops a set of stationarity and s-normality conditions for the constrained signomial problem and applies them to the reactor-cooler problem of the preceding section. It demonstrates how to compute first logarithmic derivatives for constrained signomial problems. Although these could be used in the gradient method, they will be instead combined with constrained second derivatives developed in the next section to permit application of the N-R method.

Consider a collection of $M + 1$ signomial functions $s_m(\mathbf{x})$, the subscript 0 indicating an objective function; the subscripts 1 through M, the constraint functions. Here

$$s_m(\mathbf{x}) \equiv \sum_{t=1}^{T_m} \sigma_{tm} t_{tm}(\mathbf{x}) \qquad m = 0, 1, \ldots, M$$

where the signum functions σ_{tm} are all given as either $+1$ or -1, and the terms $t_{tm}(\mathbf{x})$ are all positive power functions of $\mathbf{x}$.

$$t_{tm}(\mathbf{x}) \equiv C_{mt} \prod_{n=1}^{N} x_n^{\alpha_{mtn}}$$

The positive coefficients C_{mt} and the real exponents α_{mtn} are all known, but the positive interdependent variables $\mathbf{x}$ are not. The sign σ_0 of the objective s_0 at the stationary point sought may be unknown, although usually it is not hard to guess. As for unconstrained signomials, an incorrect hunch will be easily detected and reversed. All the constraints are presumed normalized and, most importantly, bounded *above*

$$s_m(\mathbf{x}) \leq 1 \qquad m = 1, \ldots, M$$

The reactor-cooler problem can be put into this form by introducing a new variable $x_3 \equiv x_1 - 1$ to remove the conservation term. Scaled to the design $x_1 = 2.36$, $x_2 = 0.54$, $x_3 = 1.36$, the cost becomes $0.932s_0 + 0.055$, where

$$s_0 \equiv 0.139x_1^{-0.6} + 0.075x_1^{-1} + 0.806x_2^{-0.313} + 0.119x_3^{0.6} - 0.139x_1$$

The signomial constraints are

$$s_1 \equiv 0.254x_1 + 0.746x_2 \lessapprox 1$$
$$s_2 \equiv x_1x_3^{-1} - x_3^{-1} \lessapprox 1$$

The activities and directions of these inequalities are easily confirmed by monotonicity analysis if one remembers that s_0 is to be minimized.

As for posynomial constraints, construct the lbf following by introducing nonnegative exponents $\lambda_0 \equiv 1$, $\lambda_m \geq 0$ for $m = 1, \ldots, M$.

$$s_0(\mathbf{x}) \geq \prod_{m=0}^{M} \left[s_m(\mathbf{x}) \right]^{\lambda_m} \equiv b(\mathbf{x})$$

The logarithmic first derivative of the lbf $b(\mathbf{x})$ is

$$\frac{\partial \ln b}{\partial \ln \mathbf{x}} = \sum_{m=0}^{M} \lambda_m \left(\frac{\partial \ln s_m}{\partial \ln \mathbf{x}} \right)$$

As in Sec. 5.8, the components of this N vector can be expressed in terms of positive *primal weights* $w_{mt} \equiv t_{mt}/|s_m|$:

$$\frac{\partial \ln b}{\partial \ln x_n} = \sum_{m=0}^{M} \lambda_m \sum_{t=1}^{T_m} \sigma_{mt}\alpha_{mtn}w_{mt} \qquad n = 1, \ldots, N$$

At a stationary point $\mathbf{x}_\dagger$ for b, each of these expressions must vanish by definition, giving rise to the *stationarity conditions* for constrained signomials.

$$\sum_{m=0}^{M} \sum_{t=1}^{T_m} \sigma_{mt}\alpha_{mtn}\lambda_m(w_{mt})_\dagger = 0 \qquad n = 1,\ldots,N$$

For more general analysis one can define *dual weights* ω_{mt} as any set of positive numbers satisfying the stationarity conditions and the s-normality conditions coming directly from the definitions of the $M+1$ signomials.

$$\sum_{t=1}^{T_0} \sigma_{0t}\omega_{0t} = \sigma_0 \qquad\qquad \sum_{t=1}^{T_m} \sigma_{mt}\omega_{mt} = 1$$

for $m = 1,\ldots,M$. With the dual weights, the stationarity conditions are said to be in *bilinear form*. A simpler expression comes from defining *s-dual variables* as for posynomial constraints:

$$\delta_{mt} \equiv \lambda_m\omega_{mt}$$

This produces the stationarity conditions in *linear form*.

$$\sum_{m=0}^{M} \sum_{t=1}^{T_m} \sigma_{mt}\alpha_{mtn}\delta_{mt} = 0 \qquad n = 1,\ldots,N$$

The $M+1$ bilinear s-normality conditions become a single s-*normality* condition for the objective function

$$\sum_{t=1}^{T_0} \sigma_{0t}\delta_{0t} = \sigma_0$$

and M *exponent conditions* giving the values of the unknown exponents λ_m

$$\sum_{t=1}^{T_m} \sigma_{mt}\delta_{mt} = \lambda_m \qquad m = 1,\ldots,M$$

All of the variables ω, δ, and λ must be nonnegative by definition.

In the reactor-heater example, the stationarity conditions are

$$
\begin{aligned}
x_1:&\ -0.6\omega_1 - \omega_2 & -\omega_5 + \lambda_1\omega_{11} & & +\lambda_2\omega_{21} & = 0\\
x_2:&\ \qquad\quad -0.313\omega_3 & +\lambda_1\omega_{12} & & & = 0\\
x_3:&\ \qquad\qquad\quad\ 0.6\omega_4 & & -\lambda_2\omega_{21} + \lambda_2\omega_{22} & & = 0
\end{aligned}
$$

Suppose now that the dual weights ω are set equal to the primal weights $\mathbf{w}$ at the point $(2.36, 0.54, 1.36)$. There will then be only two unknowns, λ_1 and λ_2, in the three stationarity conditions. To resolve this inconsistency, let the λ be chosen to satisfy only second and third conditions, which is equivalent to making $\partial \ln b / \partial \ln x_2$ and $\partial \ln b / \partial \ln x_3$ vanish, while allowing $\partial \ln b / \partial \ln x_1$ to be nonzero. The reader can verify that this gives

$$\lambda_1 = 0.313(0.806)/0.746(0.54) = 0.626 \quad \text{and} \quad \lambda_2 = 0.6(0.119) = 0.0715$$

This choice of exponents λ_1 and λ_2 now defines the lower bounding function completely, and one could seek the stationary point for it. Its first logarithmic derivative $\partial \ln b / \partial \ln x$, easily computed as $(0.202, 0, 0)$ at the point of departure, could be used for a gradient search, provided only one variable were changed. This is because with two constraints, there remains but one degree of freedom. In the example this would involve decreasing x_1, whose derivative is positive, in hopes of finding a stationary point where the approximating function $b(\mathbf{x})$ is less. Naturally, finding the minimum along this line of search would be entirely equivalent to doing a one-dimensional direct search in the first place, which in fact was the procedure followed in the preceding section. Instead of this, the second logarithmic derivatives will be computed for use in the N-R method, which will converge in this example.

Aside from their computational value, to be demonstrated in the section following, the stationarity and s-normality conditions can be used just like the orthogonality conditions to deduce design relationships or to generate a starting solution among the dual variables, as in the merchant fleet design problem of Chapter 4. Such preliminary analysis, however, must be followed by rigorous bounding procedures to be sure that any stationary points found are globally optimal.

6.7. SECOND LOG DERIVATIVES

The second logarithmic derivatives of the lbf $b(\mathbf{x}) = s_0 s_1^{\lambda_1} \ldots s_M^{\lambda_M}$ are easily written as linear functions of the second log derivatives of the individual signomial functions. In matrix form,

$$\frac{\partial^2 \ln b}{\partial (\ln \mathbf{x})^2} = \sum_{m=0}^{M} \lambda_m \frac{\partial^2 \ln s_m}{\partial (\ln \mathbf{x})^2}$$

A typical element of this matrix is therefore

$$\frac{\partial^2 \ln b}{(\partial \ln x_i)(\partial \ln x_j)} = \sum_{m=0}^{M} \lambda_m s_m^{-1} \frac{\partial}{\partial \ln x_i}\left(\frac{\partial s_m}{\partial \ln x_j}\right) = \sum_{m=0}^{M} \lambda_m s_m^{-1} \sum_{t=1}^{T_m} \alpha_{tj} \frac{\partial t_t}{\partial \ln x_i}$$

$$= \sum_{m=0}^{M} \sum_{t=1}^{T_m} \alpha_{ti} \alpha_{tj} \lambda_m w_{mt}$$

At a given point, the weights w_{mt} are all known, and the exponents λ_m are specified for any particular lower bounding function b. Therefore the full matrix of second logarithmic derivatives can be computed at the same point where the first log derivatives have been calculated.

In the example this computation is

$$\frac{\partial^2 \ln b}{\partial (\ln x)^2} = \begin{bmatrix} 0.36(0.139)+0.075-0.139 & 0 & 0 \\ & 0.098(0.806) & 0 \\ & & 0.36(0.119) \end{bmatrix}$$

$$+0.626 \begin{bmatrix} 0.254(2.36) & 0 & 0 \\ & 0.746(0.54) & 0 \\ & & 0 \end{bmatrix} +0.0715 \begin{bmatrix} 1.735 & 0 & -1.735 \\ & 0 & 0 \\ & & 1 \end{bmatrix}$$

$$= \begin{bmatrix} 0.485 & 0 & -0.124 \\ 0 & 0.331 & 0 \\ -0.124 & 0 & 0.114 \end{bmatrix}$$

Solving the linear N-R equations, using this matrix and the vector of first log derivatives computed in the preceding section (don't forget to change the sign), gives $\Delta \ln x_1 = -0.577$. There is no need to solve for the other two components because the problem only has one degree of freedom, making the use of all three components inconsistent with the constraints. The new value of x_1 is therefore $2.36 e^{-0.577} = 1.34$, just beyond the designs obtained by the rough direct search used in Sec. 6.5. Although the improvement will be small, it is a worthwhile exercise to perform one iteration just to see how the derivatives and exponents change (Exercise 6-7).

The N-R method could just as easily be applied to posynomials. One of the main advantages of power functions then is not so much the duality theory as it is the facility with which logarithmic derivatives can be calculated for the N-R method. For constructing lower bounds, of course, posynomials are easier to work with, and fully posynomial problems do not have any stationary points aside from the global minimum.

With signomials, second-order analysis is needed to determine the character of any stationary point found. If, in the reactor-heater example, the N-R method were started at the base case $x_1 = 1$, convergence would have been to the nearby local *maximum*, an embarrassing thing to have happen if one is not expecting it. Worse still, applying the N-R method to the reservoir problem of Sec. 6.3 would always yield the globally *worst* design, since the only stationary point there is a maximum. As the Romans have said, "*Optima corrupta pessima.*" The best things, corrupted, become the worst.

Even more insidiously, signomial problems can have several local optima. To avoid stopping at the wrong one, be sure to construct bounding functions to test for *global* optimality, either with the secant-bounding procedure of Sec. 6.5, or by constructing a lower-bounding function with zero degrees of difficulty as in Sec. 5.7. The next section shows how to extend this latter method to problems with signomial constraints.

6.8. SIGNOMIAL APPROXIMATION

The geometric inequality, so useful for identifying satisfactory designs for posynomials, is not available for signomials because of their negative terms. Since the signomial dual function may just as easily be greater than less than the primal at nonstationary points, rough lower bounds cannot be computed. But at *stationary* points, primal and dual are equal for signomials, a fact already proven in Sec. 5.9. In the present context of constrained signomials,

$$s_m(\mathbf{x}_\dagger) = \left[\prod_{t=u}^{T_m} \left(\frac{C_{mt}}{\delta_t^\dagger} \right)^{\sigma_t \delta_t^\dagger} \right]^{\sigma_m} \equiv \bar{d}_m^\dagger \qquad m = 0, 1, \ldots, M$$

where $\sigma_m \equiv 1$ for $m \geqslant 1$. Hence to define acceptable convergence to a constrained stationary point, one need only bound $|s_0 - \bar{d}_0|$, the absolute value of the difference between primal and dual objectives.

$$|s_0 - \bar{d}_0| \leqslant \tau_0$$

Extension of this simple idea to the constraint functions, which must be exactly unity to be active, requires a set of *constraint thresholds* τ_m, one for each signomial constraint.

$$\left|1-\bar{d}_m\right| \leqslant \tau_m \qquad m=1,\ldots,M$$

For posynomials, in which d_m is always a lower bound for the constraint function s_m, the difference can be used instead of the absolute value, as was done in the fleet design problem of Chapter 4. This made possible the construction of lower bounds far away from the minimum in Sec. 4.8.

To test the design $(1.70, 0.762, 0.70)$ for nearness to the global optimum, a zero-degree-of-difficulty lbf will be constructed using the primal weights for this case. The unique stationary point of this lbf will have an objective value matching that at the base case to within an acceptable tolerance, and second-order analysis will show if the point is a local, and, therefore, global minimum for the lbf.

Let the objective and constraint functions be scaled by letting $\bar{x}_1 \equiv x_1/1.70$, $\bar{x}_2 \equiv \bar{x}_2/0.762$, $\bar{x}_3 \equiv \bar{x}_3/0.70$, so that

$$y = 0.909\left[1.103\left(0.158\bar{x}_1^{-0.6} + 0.096\bar{x}_1^{-1} + 0.674\bar{x}_2^{-0.313} + 0.075\bar{x}_3^{0.6}\right)\right.$$
$$\left. - 0.103\bar{x}_1\right] + 0.055 \equiv 0.909s_0 + 0.055$$

and the constraints are

$$s_1 \equiv 0.432\bar{x}_1 + 0.568\bar{x}_2 \leq 1 \text{ and } s_2 \equiv 2.43\bar{x}_1\bar{x}_3^{-1} - 1.43\bar{x}_3^{-1} \leq 1$$

To construct the zero degree of difficulty approximating problem, first condense constraint s_1 using the usual geometric inequality and the weights at the case in question. This gives the single term inequality

$$s_1' \equiv \bar{x}_1^{0.432}\bar{x}_2^{0.568} \leq 1$$

Next condense the four positive terms of the objective in the same manner to obtain the new objective function

$$s_0' \equiv 1.103\bar{x}_1^{-0.191}\bar{x}_2^{-0.210}\bar{x}_3^{0.0446} - 0.103\bar{x}_1$$

This still leaves one degree of difficulty, since the geometric inequality cannot be used on the signomial objective or on the signomial constraint $s_2 \equiv 2.43\bar{x}_1\bar{x}_3^{-1} - 1.43\bar{x}_3^{-1} \leq 1$ remaining.

This obstacle is circumvented by using the geometric mean as an approximation rather than a lower bound. Thus transposition of the negative term gives $2.43\bar{x}_1\bar{x}_3^{-1} \leq 1 + 1.43\bar{x}_3^{-1}$, whose posynomial right member will exactly equal the geometric mean $2.43 \bar{x}_3^{0.412}$ when $x_3 = 0.70$. Division of both sides of the resulting inequality yields a new approximate constraint $s_2' \equiv \bar{x}_1 \bar{x}_3^{0.412} \leq 1$.

Recall that condensation of posynomial constraints like s_1 will in principle generate an infeasible design. But here, as in the fleet design problem of Sec. 4.9, any infeasibility not affecting the second decimal place will be regarded as numerically insignificant. The new situation here is that the approximation of the signomial constraint s_2 will in principle generate not an infeasible design, but one for which the constraint is not quite active. This extra feasibility must be controlled carefully because too much will increase the objective for the zero-degree-of-difficulty problem and destroy its value as a lower bound. The remedy is, however, the same as for the other constraint. Just neglect any infeasibility not detectable in the second decimal place.

The stationarity and s-normality conditions for the zero degree of difficulty approximation are

$$
\begin{array}{llll}
x_1: & -0.191\delta_{01} - \delta_{02} + 0.432\lambda_1 & +\lambda_2 = 0 \\
x_2: & -0.210\delta_{01} & +0.568\lambda_1 & = 0 \\
x_3: & 0.0446\delta_{01} & -0.412\lambda_2 = 0 \\
& \delta_{01} - \delta_{02} & = 1
\end{array}
$$

The unique solution is $\delta_{01} = 1.076$, $\delta_{02} = 0.076$, $\lambda_1 = 0.398$, $\lambda_2 = 0.117$, giving a dual function value of

$$
\left(\frac{1.103}{1.076} \right)^{1.076} \left(\frac{0.103}{0.076} \right)^{-0.076} = 1.004
$$

This is well within the threshold of numerical significance.

The constraints must now be checked to be sure they remained acceptably close to unity. The reader can confirm that the design obtained from the invariance conditions is $(1.06, 0.96, 1.14)$, for which $s_1 = 1.001$ and $s_2 = 0.995$, both acceptable. The point of approximation apparently was close enough to the stationary point for sufficient accuracy.

The reader can verify that the matrix of second derivatives for the lower bounding function $s_0'(s_1')^{0.398}(s_2')^{0.117}$ is

$$
\begin{bmatrix}
0.129 & 0.142 & -0.058 \\
 & 0.177 & -0.010 \\
 & & 0.032
\end{bmatrix}
$$

Application of the diagonalization method of Sec. 5.13 shows that this matrix is positive definite. Consequently, the lbf is locally minimum, and, since the stationary point is unique owing to the fact that the condensed problem has zero degrees of difficulty, it must be a *global* minimum. This independently confirms the conclusions obtained in Sec. 6.5 using secant bounds.

Although the success of this method is cause for some celebration, a certain measure of luck was involved. The matrix was almost singular, and only slight changes in the numbers could have made it indefinite. If this had happened, the analysis would have been inconclusive, making the secant bounding method the only one available. This emphasizes that although positive definiteness of the lbf is sufficient for global minimality, it is not necessary. In principle there may be global optima for which all possible zero-degree-of-difficulty lower-bounding functions are not optimal there.

6.9. AMMONIA STORAGE—A SECANT BOUND ON A DISECONOMY

This section solves a design problem with two variables, one a temperature occurring exponentially. It shows how easily the nonexponential variable can be optimized out of the problem when there are no degrees of difficulty. However, this introduces reversed inequalities in the exponential variable. Still, the remaining problem can be analyzed using what was developed in Secs. 6.1–6.5 on conservation constraints, even though the temperature involved occurs exponentially. This analysis exposes an easily overlooked local minimum.

Bounding the objective is harder here than in the previous examples. The secant bounds of Sec. 6.3 can be used on the conservation term whose exponent is less than unity, but the other, having an exponent greater than one, lends itself only to tangent bounds having too much error for conclusive results. It then becomes necessary to construct a power function *secant* bound. Since the function being approximated is the exponential part of the problem, the new lbf is a signomial. Although possessing a negative term, it fortunately has zero degrees of difficulty. Lower bounds are easily constructed, but several are needed before a conclusive bound can be obtained. One of the bounds, being in a region where a secant bound cannot be constructed, requires direct handling of an exponential objective, not a difficult task in this example.

The bounding procedures are artificial once the problem has been reduced to one with a single variable, for then a complete search is only a matter of minutes on a programmable calculator. The example should therefore be regarded only as illustrating bounding procedures useful when the presence of several variables makes direct search impractical.

The example was formulated by W. F. Stoecker in his book *Design of Thermal Systems* (McGraw-Hill, 1971, pp. 152-155) to illustrate the method of steepest descent (gradient) direct search method. Its statement follows:

An insulated steel tank storing ammonia...is equipped with a recondensation system which can control the pressure and thus the temperature of the ammonia. Two basic decisions to make in the design of this tank are the shell thickness and insulation thickness.

If the tank operates with a temperature near ambient, the pressure in the tank will be high and a thick, expensive vessel required. On the other hand, to maintain a low pressure in the tank requires more operation of the recondensation system because there will be more heat transferred from the environment unless the insulation is increased—which also adds cost.

Determine the optimum operating temperature ($t°$F) and insulation thickness (x in.) if the following costs and other data apply...

Stoecker showed the total cost to be the sum of the insulation cost $i \equiv 400x^{0.9}$, the vessel cost $v \equiv 1000 + 22(p - 14.7)^{1.2}$, where p is the pressure (psia), and the recondensation cost $r \equiv 144(80 - t)/x$. The pressure is related to the temperature by

$$\ln p = -3950(t + 460)^{-1} + 11.86$$

which is how the problem becomes exponential. By direct gradient search, the cost is found to be locally minimum (or at least stationary) at $x = 5.94$ in. (151 mm) and $t = 6.29°$F$(-14.3°$C$)$. Consider the present task then to be confirming or denying the global optimality of this design, which can serve as a base case.

Before coping with the exponential variable, notice that if the recondensation cost is considered a single positive term, then there are no degrees of difficulty in the algebraic variable x. The semilog derivative of the total cost wrt x, on being set to zero to generate an optimality condition, is

$$0.9i - r = 0$$

This gives the interesting design rule that at the optimum, the recondensation cost should be 0.9 times the insulation cost. The optimal sum of these costs can therefore be expressed as a single term by constructing the geometric mean weighted by $(1.9)^{-1} = 0.526$ and $0.9/1.9 = 0.474$.

$$i_\star + r_\star = \left(\frac{i}{0.524}\right)^{0.524} \left(\frac{r}{0.474}\right)^{0.474} = 492(80 - t)^{0.474}$$

This is the optimized sum of insulation and recondensation cost for any given temperature t. The total cost, partially optimized wrt x, now depends only upon the temperature t after elimination of the pressure.

$$c(x_\star, t) = 492(80 - t)^{0.474} + 22\left[141.5(10^3)\exp\left(\frac{-3950}{t + 460}\right) - 14.7\right]^{1.2} + 1000$$

The cost is not only exponential; it also has conservation terms that cannot be transformed into power functions without introducing reversed inequalities, as the reader can verify. Of great immediate interest is the first term, since its exponent is less than unity. This indicates, as proven in Sec. 6.1, that there is a local minimum at $80 - t = 0$, which corresponds to a system operating at ambient temperature without insulation or recondensation. It is prudent therefore to compare the cost of this ambient system with that of the base case local optimum at $6.29°F(-14.3°C)$, namely, $\$5.34(10^3)$. The ambient system turns out to be 3% less expensive—$\$5.20(10^3)$. Thus the interior local minimum found by the gradient method is not globally optimal, and at first glance it appears uneconomical to have recondensation in this application.

One can't help wondering, however, if there are not more local optima still. Since the objective now has only one variable, the simplest way to find out would be to evaluate it at several temperatures, a straightforward task for a programmable calculator. Let us instead seek to prove or disprove that the ambient design is optimal by constructing lower-bounding functions. This will suggest what must be done when there are several independent variables.

First replace the temperature with a reciprocal absolute temperature, scaled to be unity when $t = 80°F$ in the base case.

$$\xi \equiv 540(t + 460)^{-1}$$

The objective function becomes

$$c(x_\star, \xi) = 9.71(10^3)(1 - \xi^{-1})^{0.474} + 22\left[141.5(10^3)e^{-7.315\xi} - 14.7\right]^{1.2} + 10^3$$

To determine the range of the independent variable, use the fact that the lowest possible vapor pressure is 14.7 psia, which of course is 1 atm. The corresponding temperature is computed from the vapor pressure equation as $t = -29°F$, so the maximum value of ξ is 1.253: $1 \leq \xi \leq 1.253$.

Since the first term has an exponent less than unity, a secant lower-bounding function could be readily constructed for this range. But the second term, with its exponent greater than unity, would require a tangent lbf that is very inaccurate far from the base case. In this case, the bounding method which worked on the reactor-cooler design problem of Sec. 6.5 would fail, the lbf having its minimum always at the end of the range opposite from where it was constructed. Only a secant lbf, with its zero error at the ends of the range, has any hope of success.

Although the variable part of the vessel cost is a convex function of the pressure p and consequently is bounded below by any tangent, the logarithm is a concave function of both $\log p$ and $\log \xi$, as Fig. 6-8 shows. Thus a secant lbf can be constructed in the logarithmic domain that produces a power function lbf. This not only solves the bounding problem, but also circumvents the transcendentality, the new lbf being a power function in ξ.

In this case the lbf cannot be constructed over the entire range, since the term vanishes at the upper end where the vapor pressure is atmospheric.

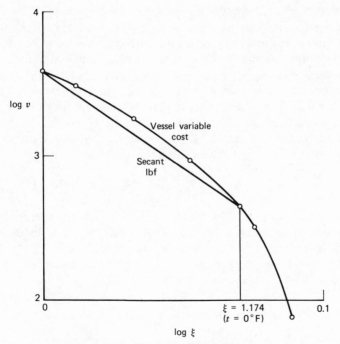

Figure 6-8 Secant bound on an exponential

Consequently the first range will be $0°F \leq t \leq 80°F$, or $1 \leq \xi \leq 1.174$. The reader can verify that the secant shown in Fig. 6-9 gives rise to the function $4.198(10^3)\xi^{-14.3}$ as the lbf for the variable part of the vessel cost. An ordinary secant lbf for the equipment cost $(i_\star + r_\star)$ is $26.54(10^3)(1 - \xi^{-1})$, so the combined lbf is the right member of

$$(10^{-3})c \geq 27.54 - 26.54\xi^{-1} + 4.198\xi^{-14.3} \equiv l_1(10^{-3})$$

Although l_1 has a negative term, it has zero degrees of difficulty, so its stationary point is easily found to be at $\xi_1 = 1.063$. Since the second semilog derivative is positive, l_1 is minimum there, having the value $4.33(10^3) < 5.20(10^3) = c(x_\star, 1)$. Hence this lower bound leaves too wide a gap, and the analysis must continue. The cost at ξ_1 is $5.65(10^3)$, showing that the apparent minimum in l_1 is probably artificial due to the large error at midrange.

In the narrower range, $1 \leqslant \xi \leqslant 1.063$, a secant lbf having no error at the ends of the range is given by the right member of

$$(10^{-3})c \geqslant 41.4 - 40.4\xi^{-1} + 4.20\xi^{-11.3} \equiv l_2(10^{-3})$$

The minimum of l_2 is $5.14(10^3)$, which is only 1.2% less than the cost of the ambient design. However, the true cost there at $\xi_2 = 1.016$ is 5.90, indicating that approximation error is again depressing the lower bound artificially. The limited accuracy of the original numbers, added to the unlikelihood that an engineer would install such a system to gain at most $50, would suggest that the ambient design be accepted, at least for temperatures above $0°F$. The iron-clad proof that the ambient design is the best is left as an exercise in lbf construction.

In the adjacent region $1.063 \leq \xi \leq 1.174$, where Stoecker found the local minimum at $\xi_\star = 1.158$, the secant lbf is

$$(10^{-3})c \geq 18.22 - 15.61\xi^{-1} + 5.789\xi^{-16.3} \equiv l_3(10^{-3})$$

This is minimum at $\xi_3 \equiv 1.125$, where $l_3 = 5.20(10^3)$, indistinguishible from the cost of the ambient design. However, the true cost there is $c_3 = 5.38(10^3)$, so the region can definitely be eliminated.

In the temperature range below $0°F$, the optimized equipment cost can be bounded below by a secant in the usual way, giving

$$9.71(10^3)(1 - \xi^{-1})^{0.474} \geq (13.75 - 11.53\xi^{-1})(10^3) \qquad 1.174 \leq \xi \leq 1.253$$

Since the variable part of the vessel cost vanishes at $\xi = 1.253$, its logarithm there cannot be used to construct a secant lbf. One can, however, make the following tangent lbf on the pressure factor

$$(p - 14.7)^{1.2} \geq 1.962p - 32.65$$

Because the range is small, the maximum error this produces in the vessel cost is only \$80, which is tolerable. When the pressure is again expressed as a function of temperature, the following lbf is generated.

$$(10^{-3})c \geq 14.03 - 11.53^{-1} + 6108e^{-7.315\xi} \equiv l_3(10^{-3})$$

The semilog derivatives, evaluated at $\xi_3 \equiv 1.174$, are

$$(10^{-3})\left(\frac{\partial l_3}{\partial \ln \xi}\right)_3 = (t_1 - 7.315\xi t_2)_3 = 9.82 - 7.315(1.174)(1.14) = 0.03 > 0$$

$$(10^{-3})\left(\frac{\partial^2 l_3}{\partial \ln \xi^2}\right)_3 = \left[-t_1 - 7.315(1 - 7.315\xi)t_2\right]_3 = 53.5 > 0.$$

Although the second derivative will change sign as ξ approaches 1.253 and t_2 vanishes, the first derivative will remain positive. Hence the lbf increases with ξ throughout the range and must have its minimum at $\xi = 1.174$, where its value, matching that of c, is \$5.35(10^3) > \$5.20(10^3) = c_0$. Since this lower bound exceeds the cost of the ambient design, the low-temperature range cannot contain a better design than the one already available.

Figure 6-9 shows all the costs and bounds computed, with the cost function and lbfs sketched in. The cost has two local minima and two local maxima, a fact easily determined with much less work by direct computation on a programmable calculator. The example was solved as it was to illustrate the construction of bounding functions and the handling of an exponential variable.

Increasing the vessel cost only 10% would make the recondensation scheme attractive. Analysis of this situation is left as an exercise of special interest to designers of refrigeration machinery.

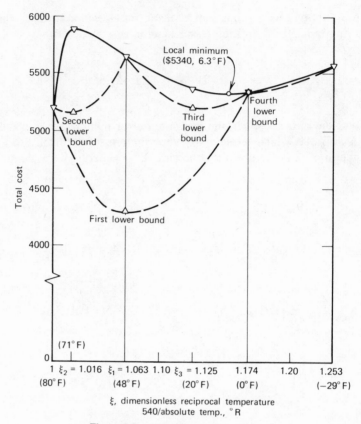

Figure 6-9 Ammonia storage system

6.10. CONCLUDING SUMMARY

The conservation laws and shifts of origin so common to engineering complicate the designer's task by generating multiple stationary points and making inapplicable the elegant bounding procedures of geometric programming. Faced with an active reversed inequality, the designer is advised to try using it to eliminate a variable, even though this usually destroys any signomial structure. If the variable eliminated can go to zero, often the case when a conservation law is involved, and if its exponent is less than unity, then there must be a local optimum just where it vanishes. For the equipment associated with such a variable to be justified at all, it would therefore have to produce another local optimum that is even better. The designer's nightmare is that maybe it won't.

Since having an exponent less than unity arises here from economy of scale, this situation is far from rare in practice. The extra component—a reservoir, a heat exchanger, or a refrigeration compressor—does not pay its way when small, forcing the system to go through an unwanted pessimum uncomfortably close to what may be the global optimum. Before looking for the local optimum associated with the auxiliary component, however, the designer should check the simpler case where the component is absent. Constructing a lower bound there might prove the auxiliary to be uneconomical anyway. And when relying on a vendor to design an auxiliary component, an engineer should at least compare its cost to that of the simpler system to see if the extra sophistication is really economical.

Fortunately, the secant and tangent bounds needed to ensure global optimality are not much harder to construct than they were for posynomials. Usually, however, the error between points of approximation artificially enlarges the gap between design cost and its lower bound, making necessary the construction of bounds in several separate regions. This takes more work than before, but of course the problem is harder.

Once satisfied that the auxiliary pays its way, the designer can put the problem in signomial form during the search for the nearest local, hopefully global, optimum. Although bounds are not available early in the search, at least the condensation and partial invariance methods can improve base cases. And when there are many degrees of difficulty, the easily computed logarithmic first and second derivatives make practical the N-R method, suitably modified for constraints. Of course, this procedure must be watched carefully, using second-order information, to keep from homing in on a saddle or local pessimum.

When lack of further improvement indicates possible proximity to an optimum, at least a local one, signomial approximation can be used to generate a lower bound. Close control of any signomial constraints is essential, for any substantial error will make the lower bound artificially high, risking premature termination away from the true optimum. As in Chapter 4, posynomial constraint error must also be small to retain feasibility. With luck, this single bound construction will positively identify a global optimum, although approximation error may require bounds in several smaller regions to prove optimality conclusively.

To keep things simple, the examples, after preliminary partial optimization, involved but one degree of freedom, so they could have been solved by direct exhaustive search on a pocket calculator. Despite this, the original problems were far from trivial, being capable of thoroughly confusing most existing nonlinear programming codes. Of some engineering interest in their own right, they are vehicles for demonstrating how to handle the negative terms without missing the global optimum.

The ammonia storage example incidentally involved a temperature occurring exponentially. The next chapter gives more general methods for solving such "transcendental" design problems.

NOTES AND REFERENCES

Secant bounds were first used by Barry McNeill in solving the pipeline-reservoir problem. They also appeared in the *AIChE Journal* reactor cooler article.

Gary Blau used the signomial structure to do a N-R solution of a batch reactor design.

Research-minded designers are referred to the articles by Passy and Rijckaert applying the "equilibrium conditions," not discussed in this book, to signomial reactor design.

Blau, G. E., and D. J. Wilde, "Optimal System Design by Generalized Polynomial Programming," *Can. J. Chem. Eng.*, **47**, 317–326 (1969).

Passy, U., and D. J. Wilde, "Mass Action and Polynomial Optimization," *J. Eng. Math.*, **3**, 325–335 (1969).

Rijckaert, M. J., "Engineering Applications of Geometric Programming," in Avriel, Rijckaert, and Wilde, pp. 196–220, (*op. cit.* Chapter 1).

Wilde, D. J., "Global minimization of a Cooled Reactor using a Posynomial Lower Bounding Function," *Amer. Inst. Chem. Engrs. J.*, **22**, 4 (1976) 685.

Wilde, D. J. and B. W. McNeill, "Economic Design of a Pipeline with Discharge Reservoir," *Engng. Optzn.* (To appear.)

EXERCISES

*6-1. Examine your design project for conservation constraints and reversed inequalities. Use each of them to eliminate a variable in such a way that locally optimal designs become apparent. List any locally optimal designs made obvious in this way and discuss the engineering significance of such designs (Sec. 6.1).

6-2. Design a pipeline and reservoir system in which the objective is the same as in Sec. 6.3, except that the storage cost coefficient is twice as high because of unfavorable topography.

6-3. Design a pipeline and reservoir system in which the objective is the same as in Sec. 6.4, except that the storage cost coefficient is twice as high because of unfavorable topography.

6-4. Suppose that increased energy costs in the chemical reaction system of Sec. 6.5 double the energy cost and steam credit terms. What is the maximum fixed cost that can be justified for an auxiliary cooler?

6-5. Verify the results in Sec. 6.5 for: (a) $1.00 \leq x_1 \leq 1.40$, (b) $2.00 \leq x_1 \leq 3.94$, (c) $1.40 \leq x_1 \leq 2.00$.

*Project problem.

*6-6.   Apply the bounding methods of Secs. 6.3–6.5 to your project, if it has conservation constraints or reversed inequalities.

6-7.   Perform another iteration on the example in Sec. 6.7.

6-8.   Write the pipeline-reservoir problem of Sec. 6.4 in the signomial form of Sec. 6.6. Write the s-normality and stationarity conditions in both bilinear and linear form. Compute first and second logarithmic derivatives at the base case ($q = 1$) and make one application of the N-R method (Sec. 6.7).

6-9.   For Problem 6-8, construct a lower bound by the signomial approximation method of Sec. 6.8.

*6-10.   If your project has conservation constraints or reversed inequalities, apply the N-R method of Secs. 6.6 and 6.7 to it, finishing with a bound constructed by the signomial approximation method of Sec. 6.9.

6-11.   In the ammonia storage example of Sec. 6.9, increase the storage tank cost 10% and find the minimum cost design.

CHAPTER 7

Transcendental Thermal and Diffusion Systems

Duality is the fundamental cause of suffering.

Maharishi Mahesh Yogi (c. 1970)

The preceding chapter coped with shifts of origin, often caused by having temperature as one of the variables. This chapter studies another temperature effect. When chemical reaction or, as in the ammonia storage problem of Sec. 6.9, phase equilibrium is involved, the physical relations cannot be expressed precisely as power functions of the temperature. When this happens, there is no way to make the temperature cancel from bounds constructed by the geometric inequality.

Such functions, being exponential, transcend previous theory in the Aristotelian sense of reaching beyond the bounds of any category. Mathematicians therefore call functions not expressible by the operations of addition, subtraction, multiplication, or division by the name *transcendental*. Logarithmic, trigonometric, and inverse trigonometric functions are all transcendental.

Only exponential and logarithmic functions will be studied here. Trigonometric functions can be transformed into signomials by using rectangular instead of polar coordinates. The theory for the inverse trigonometric functions such as arctangent is not included.

Potential applications of this theory, naturally called transcendental programming, reach especially into the design of processing systems involving transport of energy and mass. Temperature is exponential in chemical reaction rates and vapor pressures, while multistage separation processes involve logarithmic functions of concentration differences. A gaseous diffusion plant design example will illustrate this latter situation.

Many methods for handling power functions by geometric programming still apply to the algebraic variables—all those except the transcendentals. Stationarity conditions for the transcendental variables are unusual in involving transcendental as well as the customary dual variables. But since each relation involves only one transcendental, and that bilinearly, such problems can be solved completely when there are no degrees of difficulty for the algebraic variables. Condensation and partial invariance procedures can be applied to some problems with algebraic degrees of difficulty, and occasionally lower bounds can be constructed. On the other hand, the complexity of the second derivatives precludes the N-R method.

Being the newest branch of optimization in design, transcendental programming has not yet seen wide application. It should, however, quickly bring optimization to bear on some of the most difficult features of process and thermal design problems.

7.1. OPTIMAL REACTION TEMPERATURE

This chapter deals with transcendental programming—optimization problems with unavoidable exponential or trigonometric terms. For example, an exponential cost term of the form e^{ar} can be transformed into a power function x^α by letting $x \equiv e^r$, but this does not work if r also appears in a power function, as in the term $r^\beta e^{\gamma r}$.

Unavoidable exponential or trigonometric terms, henceforth called transcendental, occur frequently in processes involving temperature as a design variable. Reaction rates, transport properties such as viscosity, and physical properties such as vapor pressure, can depend exponentially on absolute temperature, whereas heat-transfer rates tend to be at worse power functions of temperature. A grossly oversimplified example illustrates this.

Consider a chemical reaction system in which the reaction is allowed to approach equilibrium at the operating temperature T (absolute, in °K), to be selected by the process designer. If the reaction requires heat, in which case it is called *endothermic*, the Second Law of Thermodynamics predicts that conversion is improved at higher temperatures. Moreover, the reactor can be smaller at high temperatures because of the increased velocity of reaction. The associated costs depend exponentially on the reciprocal of the absolute temperature, being proportional to $e^{\gamma/T}$, with γ a known parameter, positive for endothermic reactions so that the factor decreases wrt T. On the other hand, achieving the higher temperatures is expensive, requiring both energy and equipment, so assume these costs proportional to T^β, with β positive. Although a cost function of the form $T^\beta + Ke^{\gamma/T}$, with K a parameter, would qualify as an untransformable exponential, i.e., transcendental, function, consider instead the single term $T^\beta e^{\gamma/T}$ obtained by multiplying the two factors together. The former, more realistic, function will be treated in later sections; the latter serves as the simplest possible transcendental function for introducing the subject of transcendental optimization.

It is convenient for differentiation to make the exponent of e linear, so let $r \equiv T^{-1}$ be the reciprocal of the absolute temperature, making the cost function become $r^{-\beta}e^{\gamma r} \equiv y$. The first semilog derivative is

$$\frac{\partial y}{\partial \ln r} = r(-\beta r^{-(\beta+1)}e^{\gamma r} + \gamma r^{-\beta}e^{\gamma r}) = (-\beta + \gamma r)y$$

Unlike the semilog derivative for a power function, this depends on the variable r. Fortunately the dependence is linear, so the derivative can be set to zero and solved for the unique stationary point:

$$r_\dagger = \frac{\beta}{\gamma}$$

The second semilog derivative is

$$\frac{\partial^2 y}{\partial (\ln r)^2} = (-\beta + \gamma r)\frac{\partial y}{\partial \ln r} + \gamma\left(\frac{\partial r}{\partial \ln r}\right)y$$

At the stationary point r the first term vanishes, leaving only

$$\left(\frac{\partial^2 y}{\partial (\ln r)^2}\right)_\dagger = \gamma r_\dagger y_\dagger = \beta y_\dagger > 0$$

Thus $r_\dagger$ is the global minimum and can be labelled $r_\star$.

To gain an idea of scale, let $\beta = 4$ and $\gamma = 5000$, numbers faintly reasonable for some weakly endothermic processes. Then $r_\star = 4/5000$, giving an optimal temperature $T_\star = 1250°K$ (977°C). Most endothermic reactions have larger exponents γ, indicating "optimal" temperatures impractically high. Such processes simply operate at the maximum possible temperature, as in the system of the chapter preceding.

Exothermic reactions, which give off rather than absorb heat, have negative exponential coefficients γ, making $r^{-\beta}e^{\gamma r}$ monotonically decreasing in r, or increasing in T, provided β remains positive. More realistically,

β should also be negative for exothermic reactions, since the coolers required, as well as the reactor, cost more at low temperatures. With both β and γ negative, such a system would have its second derivative negative at the unique stationary point, indicating a globally *maximum* cost there. If this point were inside the allowable temperature range, there would be a local minimum at each extreme. The maximum is more likely to be above the range, in which case the minimum feasible temperature would be best. But this oversimplified example should be regarded merely as an introduction to the optimization theory of transcendental functions rather than a hard and fast study of chemical reactor design. A more realistic process problem is solved in Sec. 7.5 after the theory needed has been derived.

7.2. TRANSCENDENTAL FUNCTIONS

Any variable occurring in exponential as well as power functions will be called *transcendental* and symbolized by ξ (*xi*) to contrast it with the other variables $\mathbf{x}$, called *algebraic*, which appear only in power functions. Transcendental problems, although harder than purely algebraic ones, are not as difficult as they might seem at first glance. The bounding methods already developed can be applied directly to the algebraic variables, leaving a transcendental residue to be handled by techniques to be developed now.

Consider a typical transcendental term, with its index omitted for simplicity.

$$t \equiv C \prod_{n=1}^{N} x_n^{\alpha_n} \prod_{p=1}^{P} \xi_p^{\beta_p} \exp(\gamma_p \xi_p)$$

Here the N algebraic and P transcendental variables are assumed positive, and the exponents are known real numbers. The semilog derivative wrt any algebraic variable, say x_i, is the same whether there are transcendental variables or not: $\partial t / \partial \ln x_i = \alpha_i t$. With respect to a typical transcendental variable, say ξ_j, the semilog derivative is only mildly different.

$$\frac{\partial t}{\partial \ln \xi_j} = (\beta_j + \gamma_j \xi_j) t \qquad j = 1, \dots, P$$

Thus the stationarity conditions, written as functions of the terms, will depend only on the transcendental variables, not on the algebraic ones. But the stationarity condition for a given transcendental variable involves only that variable, and in a linear fashion.

Now consider a transcendental example with four algebraic variables, two transcendentals, a three-term objective, and a two-term tight constraint. Lidor, who made up this problem, calls such functions *posynentials* by analogy with posynomials.

$$\min_{x,\xi} y(x,\xi): \quad y = 2x_1x_4 + x_1^{-1}x_2\xi_1^{-1}\xi_2^{-1}e^{-\xi_1} + x_2^{-4}x_3x_4^{-1}\xi_1^{-1}e^{3\xi_1-\xi_2}$$

$$\text{s.t.} \quad x_1x_2\xi_1 + x_2x_3^{-1}\xi_2e^{2\xi_2} \leq 1$$

For the four algebraic variables, the stationarity conditions are, as functions of dual variables defined in the usual way,

$$\begin{array}{llll}
x_1: & \delta_{01} - \delta_{02} & + \delta_{11} & = 0 \\
x_2: & \delta_{02} - 4\delta_{03} + \delta_{11} + \delta_{12} & = 0 \\
x_3: & \delta_{03} & - \delta_{12} & = 0 \\
x_4: & \delta_{01} & - \delta_{03} & = 0
\end{array}$$

In addition, the normality condition is

$$\delta_{01} + \delta_{02} + \delta_{03} = 1$$

There being exactly five independent linear equations in the five unknown components of $\boldsymbol{\delta}$, the solution is unique.

$$\boldsymbol{\delta} = \left(\tfrac{1}{4}, \tfrac{1}{2}, \tfrac{1}{4}, \tfrac{1}{4}, \tfrac{1}{4}\right)^T$$

Notice that in general the number of degrees of difficulty, zero in the example, is obtained in the usual way, the transcendental variables being ignored.

The stationarity conditions for the transcendental variables are

$$\begin{array}{lll}
\xi_1: & (-1-\xi_1)_+\delta_{02} + (-1+3\xi_1)_+\delta_{03} + \delta_{11} & = 0 \\
\xi_2: & -\delta_{02} - (\xi_2)_+\delta_{03} + (1+2\xi_2)_+\delta_{12} & = 0
\end{array}$$

But since the dual variables are known, the first becomes

$$\tfrac{1}{2}(-1-\xi_1)_+ + \tfrac{1}{4}(-1+3\xi_1)_+ + \tfrac{1}{4} = 0$$

from which $(\xi_1)_+ = 2$. The second transcendental stationarity condition is

$$-\tfrac{1}{2} - \tfrac{1}{4}(\xi_2)_+ + \tfrac{1}{4}(1+2\xi_2)_+ = 0$$

from which $(\xi_2)_+ = 1$.

The geometric inequality can now be used to construct a lbf $l(\xi)$ that is a function of the transcendental variables ξ, but totally independent of the algebraic variables **x**.

$$y(\mathbf{x},\xi) \geq \left(\frac{2}{\frac{1}{4}}\right)^{1/4} \left[\frac{\xi_1^{-1}\xi_2^{-1}e^{-\xi_1}}{\frac{1}{2}}\right]^{1/2} \left[\frac{\xi_1^{-1}e^{3\xi_1-\xi_2}}{\frac{1}{4}}\right]^{1/4} \left(\frac{\xi_1}{\frac{1}{2}}\right)^{1/4} \left[\frac{\xi_2 e^{2\xi_2}}{\frac{1}{2}}\right]^{1/4}$$

$$= 4.76\xi_1^{-1/2}\xi_2^{-1/4}\exp\left(\tfrac{1}{4}\xi_1 + \tfrac{1}{4}\xi_2\right) \equiv l(\xi)$$

It has already been shown that $y(\mathbf{x},\xi)$ is stationary at $(\xi)_\dagger = (2,1)$. Because of the geometric inequality, this point is minimum wrt **x**, but its curvature wrt ξ must be determined by examining second semilog derivatives wrt ξ. This can be deferred to Sec. 7.5, however, because in this example $(\xi)_\dagger$ must be unique, there being zero degrees of difficulty. That is, ξ cannot vary from $(\xi)_\dagger$ without violating the normality, orthogonality, and transcendental stationarity conditions unless there are degrees of difficulty. In the present example, then, $l(\xi)$ is a constant, equal to $4.76(2)^{-1/2}e^{3/4} = 7.13$. Thus $y(\mathbf{x},\xi) \geq l(\xi) = 7.13$, and since the value of the objective at $\xi = (2,1)^T$ is also 7.13, the corresponding design is the global minimum, and $\xi_\dagger = \xi_\star$.

Now that the optimal transcendental variables are known, the minimizing values $\mathbf{x}_\star$ of the algebraic variables can be found from the invariance conditions in the usual way. The invariance conditions are:

$$2(x_1 x_4)_\star = \tfrac{1}{4}(7.13) \qquad \tfrac{1}{2}e^{-2}\left(x_1^{-1}x_2\right)_\star = \tfrac{1}{2}(7.13)$$

$$\tfrac{1}{2}e^5\left(x_2^{-4}x_3 x_4^{-1}\right)_\star = \tfrac{1}{4}(7.13) \qquad 2(x_1 x_2)_\star = \tfrac{1}{2}; \quad e^2\left(x_2 x_3^{-1}\right)_\star = \tfrac{1}{2}$$

Any four of these equations can be used to find $\mathbf{x}_\star$, either by directly multiplying powers of the equations together or by taking logarithms and solving the resulting linear equations. The solution is

$$\mathbf{x}_\star = (0.0689, 3.63, 53.6, 12.9)^T$$

This example has shown how to solve a posynential programming problem having zero algebraic degrees of difficulty. After the algebraic variables have been handled in the usual way, a lbf depending on the unique values of transcendental variables remains. When there are degrees of difficulty, the methods of Chapters 3 and 4 must be employed to generate an approximate problem with zero degrees of difficulty. This will lead to a lower bound that, if close enough to the objective value for the best design available, will terminate the search. If there are negative terms, the global optimization techniques of Chapters 5 and 6 must be used.

Thus, although transcendental variables must be treated specially, the other variables can be analyzed almost as if no transcendentals were present, provided there are zero degrees of difficulty. The mathematics will be worked out rigorously in Sec. 7.4, after a section showing how to transform other types of transcendental function into posynential form.

7.3. TRANSCENDENTAL TRANSFORMATIONS

Here are four examples, developed by Lidor, of nonposynential forms and how to transform them into standard posynentials. The original physically based forms of transcendental design problems usually take one of these forms rather than the posynential one needed for optimization. All variables are assumed positive unless otherwise restricted.

1. Log-log curves. A cornerstone for geometric programming modeling has been that any function giving a straight line on a log-log plot can be represented algebraically by a power function. But when the log-log plot is a curve, the function $t(x)$ is transcendental instead. If the curve can be described by a posynomial $p(x)$ so that $t(x) = x^{p(x)}$, then it can be transformed into posynential form by the following manipulations.

$$t = x^{p(x)} = (e^{\ln x})^{p(x)} = \exp\left[p(x)\ln x \right]$$

Define the three transcendental variables:

$$\xi_1 \equiv \ln x \qquad \xi_2 \equiv p(x) \qquad \xi_3 \equiv \xi_1 \xi_2$$

Then $t = e^{\xi_3}$. If t is an objective function to be minimized or a constraint function bounded above, then since t increases in ξ_3, the third transformation is equivalent to the active inequality $\xi_3 \geq \xi_1 \xi_2$, from which

$$\xi_1 \xi_2 \xi_3^{-1} \leq 1$$

This bounds ξ_2 above, so to maintain strict equality, the second transformation must bound ξ_2 below, i.e., $\xi_2 \geq p(x)$, when

$$\xi_2^{-1} p(x) \leq 1$$

Similarly, the first transformation must bound ξ_1 below, i.e., $\xi_1 \geq \ln x$. Taking exponentials and rearranging gives

$$xe^{-\xi_1} \leq 1$$

This type of transformation can also be carried out when the log-log curve is not posynomial, or when t is to be maximized or is a constraint function bounded below. However, the result will not be posynential and must be optimized by combining the condensation and partial invariance methods.

2. Logarithmic functions. Heat and mass transport problems often have logarithmic functions, which fortunately are easily put into posynential form. Suppose

$$t(\mathbf{u}) \equiv C \prod_{r=1}^{R} (\ln u_r)^{\beta_r} u_r^{\gamma_r}$$

with $u_r > 1$ for all $r = 1, \ldots, R$. Let $\xi_r \equiv \ln u_r$ so that $u_r = e^{\xi_r}$. Then $t(\xi)$ is posynential.

$$t(\xi) \equiv C \prod_{r=1}^{R} \xi_r^{\beta_r} \exp(\gamma_r \xi_r)$$

3. Posynomial exponent. A curved semilog plot may give a term with a posynomial $p(\mathbf{x})$ in the exponent, for example,

$$t(\mathbf{x}) \equiv \exp\left[p(\mathbf{x}) \right] \prod_{n=1}^{N} x_n^{\alpha_n}$$

Let $\xi \equiv p(\mathbf{x})$. Then if t is to be minimized, or if it is a constraint function bounded above, it can be written as a posynential

$$t(\mathbf{x}, \xi) = e^{\xi} \prod_{n=1}^{N} x_n^{\alpha_n}$$

subject to the new constraint

$$\xi^{-1} p(\mathbf{x}) \leqq 1$$

4. Semilog regression. When an empirical design relation is obtained by a linear regression in several logarithmic variables, the objective can be of the form

$$y(\mathbf{x}) \equiv p(\mathbf{x}) \sum_{n=1}^{N} \alpha_n \ln x_n = p(\mathbf{x}) \ln \prod_{n=1}^{N} x_n^{\alpha_n}$$

If $\sum_{n=1}^{N} \alpha_n \ln x_n \geq 0$, then the transformation $\xi \equiv \ln \prod_{n=1}^{N} x_n^{\alpha_n}$ gives

$$y(x,\xi) = \xi p(x)$$

with the new constraints

$$e^{-\xi} \prod_{n=1}^{N} x_n^{\alpha_n} \leqq 1$$

and

$$\prod_{n=1}^{N} x_n^{-\alpha_n} \leq 1$$

These four cases cover many, but not all, transcendental situations. The restrictions are needed to give posynentials whose analysis will lead easily to the global optimum. When these restrictions do not hold, the same transformations give nonposynentials still capable of optimization, but with which precautions against multimodality or unboundedness must be taken.

7.4. TRANSCENDENTAL PROGRAMMING

This section proves the theory developed informally in Sec. 7.2 to solve the zero-degree-of-difficulty example there. Transcendental programming is shown to be a natural extension of the geometric programming theory of Secs. 3.10 and 4.2 to the optimization of posynentials.

Let $\mathbf{x}$ be an N vector and $\boldsymbol{\xi}$ an R vector of nonnegative variables, the former called *algebraic* and the latter *transcendental*. Consider $M+1$ *posynential* functions defined by

$$q_m(\mathbf{x},\boldsymbol{\xi}) \equiv \sum_{t=1}^{T_m} C_{mt} \prod_{n=1}^{N} x_n^{\alpha_{mtr}} \prod_{r=1}^{R} \xi_r^{\beta_{mtr}} \exp(\gamma_{mtr}\xi_r) \qquad m=0,1,\ldots,M$$

The *transcendental programming problem* is

$$\min_{\mathbf{x},\boldsymbol{\xi}} q_0(\mathbf{x},\boldsymbol{\xi})$$

subject to

$$q_m(\mathbf{x}, \boldsymbol{\xi}) \leqq 1 \qquad m = 1, \ldots, M$$

Let a positive *dual variable* $\delta_{mt} > 0$ be defined for every term of the problem. For $m = 0$ (the objective function), these must satisfy a *normality condition*:

$$\sum_{t=1}^{T_0} \delta_{0t} = 1$$

In addition, they must satisfy N *algebraic orthogonality conditions*, one for each algebraic variable.

$$x_n: \quad \sum_{m=0}^{M} \sum_{t=1}^{T_m} \alpha_{mtn} \delta_{mt} = 0 \qquad n = 1, \ldots, N$$

There are also R *transcendental stationarity conditions*, one for each transcendental variable.

$$\xi_r: \quad \sum_{m=0}^{M} \sum_{t=1}^{T_m} (\beta_{mtr} + \gamma_{mtr}\xi_r)\delta_{mt} = 0 \qquad r = 1, \ldots, R$$

The left members of the algebraic orthogonality and transcendental stationarity conditions are the semilog derivatives of the Lagrangian wrt $\mathbf{x}$ and $\boldsymbol{\xi}$.

Using the geometric inequality as in geometric programming with the $\boldsymbol{\delta}$ as weights gives

$$q_0 \geq \prod_{m=0}^{M} \prod_{t=1}^{T_m} \left(\frac{C_{mt}}{\delta_{mt}/\lambda_m} \right)^{\delta_{mt}} \prod_{r=1}^{R} \left[\xi_r^{\beta_{mtr}} \exp(\gamma_{mtr}\xi_r) \right]^{\delta_{mt}}$$

where $\lambda_m \equiv \sum_{t=1}^{T_m} \delta_{mt}$ for $m = 1, \ldots, M$, and $\lambda_0 \equiv 1$. As in geometric programming, the algebraic variables $\mathbf{x}$ cancel out of the lower bounding function, but the transcendentals $\boldsymbol{\xi}$ remain. Then the abbreviations

$$d \equiv \prod_{m=0}^{M} \prod_{t=1}^{T_m} \left(\frac{C_{mt}}{\delta_{mt}} \right)^{\delta_{mt}} (\lambda_m)^{\lambda_m}$$

$$\zeta_r \equiv \sum_{m=0}^{M} \sum_{t=1}^{T_m} \delta_{mt}\beta_{mtr} \qquad \eta_r \equiv \sum_{m=0}^{M} \sum_{t=1}^{T_m} \delta_{mt}\gamma_{mtr} \qquad r = 1, \ldots, R$$

give the inequality

$$q_0 \geq d \prod_{r=1}^{R} \xi_r^{\xi_r} \exp(\eta_r \xi_r) \equiv l(\xi)$$

When there are zero algebraic degrees of difficulty there is a unique solution to the normality, orthogonality, and transcendental stationarity conditions. This makes $l(\xi)$ a constant, as in the example of Sec. 7.2. When there are degrees of difficulty, as in the gaseous diffusion separation problem solved in the next section, the lower-bounding methods for solving posynomial problems are applicable to the posynential case.

7.5.   DEGREES OF DIFFICULTY IN A GASEOUS DIFFUSION PLANT

Not all posynential problems can be condensed to zero degrees of difficulty. This section will set up and solve such a problem arising from the technology of uranium isotope separation by gaseous diffusion. From base case weights a lbf is generated that depends on two transcendental variables, although not on the two algebraic variables of the problem. This lbf, evaluated where the transcendental variables satisfy normality, orthogonality, and transcendental stationarity conditions, is a valid lower bound on the original primal objective only at the point of approximation. The gap between primal cost and lower bound exceeds the 1% satisfaction threshold, so a new design is generated by the partial invariance method of Section 4.10. An iteration generates a third design that passes the satisfaction test.

The physical problem arises from the chemical engineering of uranium isotope separation by the gaseous diffusion process. Uranium hexafluoride (UF_6) gas with 0.71 mol% in the desired U^{235} fissionable isotope form is pumped through hundreds of diffusion chambers that concentrate it eventually to 3.3 mol%. At the other end of the process a waste stream of very low concentration is removed. The mol fraction x in the waste is to be chosen by trading off the cost of the U^{235} lost against the cost of achieving a sharp separation. To avoid signomials in this simple example, let x be neglected relative to the product mol fraction 0.033, as well as in the expressions $1 - x$ and $1 - 2x$ which would arise in an exact derivation. Then by mass balance, the U^{235} lost in the waste is proportional to $(0.0071 - x)^{-1}$. The associated cost must be added to the cost of pumping the gases through the diffusion chambers. The variable part of this latter cost for the *separative work* is proportional to

$$(0.0330 - 0.0071)(0.0071 - w)^{-1} \ln(w^{-1}) + \ln(0.0071 - w)^{-1}$$

Here the rigorous thermodynamic expressions have been simplified by neglecting w in the expressions $(0.033 - w)$, $1 - w$, and $1 - 2w$. An objective function is now constructed by adding these expressions, each multiplied by appropriate coefficients reflecting the costs of U^{235} and of the separative work. This variable cost v is

$$v \equiv 0.775(0.0071 - w)^{-1} + 0.743(0.0071 - w)^{-1} \ln(w^{-1})$$
$$+ 28.3 \ln\left[(0.0071 - w)^{-1}\right]$$

The coefficients reflect 1970 prices of uranium and energy.

The easiest and safest way to minimize this expression is to search directly on a programmable calculator, there being only one independent variable. But to illustrate the application of partial invariance to transcendental programming, the problem will be put in posynential form. The reader can verify that the appropriate transformations, arranged to preserve boundedness, are

$$u \leqq 0.0071 - w, \quad \xi_1 \geqq \ln(w^{-1}), \text{ and } \xi_2 \geqq \ln\left[(0.0071/w)^{-1}\right].$$

Then in standard posynential form the problem becomes to minimize

$$v = 0.775u^{-1} + 0.743u^{-1}\xi_1 + 28.3\xi_2$$

subject to $u, w, \xi_1, \xi_2 > 0$ and

$$140.8w + 140.8u \leqq 1 \qquad w^{-1}\exp(-\xi_1) \leqq 1 \qquad u^{-1}\exp(-\xi_2) \leqq 1$$

With seven terms and four variables, two of them algebraic and two transcendental, there are $7 - 2 - 1 = 4$ algebraic degrees of difficulty. Even if both posynomials were condensed all the way to one term each, there would still remain one algebraic degree of difficulty, so condensation is not available for solving the problem. Hence the partial invariance method will be used.

The orthogonality, transcendental stationarity, and normality conditions are

$$
\begin{array}{llll}
w: & \delta_{11} & -\lambda_2 & = 0 \\
u: & -\delta_{01} - \delta_{02} & +\delta_{12} & -\lambda_3 = 0 \\
\xi_1: & \delta_{02} & -\xi_1\lambda_2 & = 0 \\
\xi_2: & \delta_{03} & -\xi_2\lambda_3 & = 0 \\
& \delta_{01} + \delta_{02} + \delta_{03} & & = 1
\end{array}
$$

There is no obvious choice for any of the dual variables, but it is easy to generate a base case simply by guessing w, since there is only one degree of freedom. The physically meaningful range for w is $0 < w < 0.0071$, so take the midpoint, rounded down to one significant figure: $w_0 = 0.0030$. Then the values of the other variables in this base case are quickly determined as $u_0 = 0.0041$, $\xi_{10} = 5.809$, and $\xi_{20} = 5.49$. They determine a cost of 1397 $(0.135 + 0.753 + 0.111)$, so let the dual variables for the objective equal these primal weights. If the transcendental variables in the dual equations are also assigned their primal values, then the remaining dual variables must be fixed as $\lambda_2 = \delta_{11} = 0.753/5.809 = 0.130$, $\lambda_3 = 0.111/5.496 = 0.0203$, from which can be computed $\delta_{12} = 0.135 + 0.753 + 0.025 = 0.909$. The following lbf, expressed as a function of the transcendental variables, is then constructed.

$$v(\xi_{10}, \xi_{20}) \ge \left(\frac{0.775}{0.135} \right)^{0.135} \left(\frac{0.743}{0.753} \xi_{10} \right)^{0.753} \left(\frac{28.3}{0.111} \xi_{20} \right)^{0.111} \left(\frac{140.8}{0.125} \right)^{0.13} \left(\frac{140.8}{0.875} \right)^{0.909}$$

$$\times \left[\exp(-\xi_{10}) \right]^{0.130} \exp(-\xi_{20})^{0.0203} \equiv l(\xi_{10}, \xi_{20}) = 1121$$

This lower bound is 20% less than the base cost of 1397, so a new design must be generated.

The first term of the first constraint gives a new design immediately: $140.8w = 0.125$, from which $w = 0.0009$. The cost of this design is 1109, even less than the lower bound 1121 just generated at the base case. Computation of a lower bound for the new design, left to the reader as Exercise 7-7, gives 1107, very close to the primal cost for this design. But since this value is not a lower bound over the whole range of w, a third design $w = 0.0008$ must be generated. Its cost also being 1107, this design is accepted.

7.6. CONCLUDING SUMMARY

This chapter transcends the limitations of power-function-oriented methods developed in earlier chapters. The transcendental variables are precisely those that cannot be made to disappear by constructing a geometric mean lbf. But lower bounds can be obtained in the usual way, as in the gaseous diffusion plant example, and partial invariance is still applicable.

This concludes the study of the nonlinearities of steady-state engineering-design optimization. They were not handled by the usual linearization approach for, as the Old Testament tells us, "That which is crooked cannot be made straight" (Ecclesiastes 1:14). Instead, bounds are constructed, usually from the geometric inequality, that prevent excessive computation. This same bounding approach, using different theory, overcomes the last difficulty in the next chapter, i.e., indivisibility of the design variables.

NOTES AND REFERENCES

Duffin, Peterson, and Zener (*op. cit.* Chapter 1) approximate logarithmic and exponential functions by the limiting equations

$$\ln x = \lim_{\epsilon \to 0} (x^{\epsilon} \epsilon^{-1} - \epsilon^{-1})$$

and

$$e^x = \lim_{\epsilon \to 0} \left[(1 - \epsilon x)^{1/\epsilon} \right]$$

Numerical solutions are obtained by any of the techniques described in earlier chapters for suitably small values of ϵ. Beightler and Phillips (*op. cit.* Chapter 1) give additional examples of this method.

The material in Secs. 7.2–7.4 comes from Lidor's Ph.D. dissertation.

The physical relations for the gaseous diffusion problem are from Benedict and Pigford.

Benedict, M., and T. H. Pigford, *Nuclear Chemical Engineering*, McGraw-Hill, New York, 1957.

Lidor, G., and D. J. Wilde, "Transcendental Geometric Programs", *J. Optzn. Theory & Appl.*, 25, 1 (1978).

EXERCISES

*7-1.   If applicable, use the transformations of Sec. 7.3 to convert your project into the posynential form of Sec. 7.4.

*7-2.   If your project is posynential, write the normality, algebraic orthogonality, and transcendental stationarity conditions. Are any simple design rules apparent?

7-3.   Certain terms were neglected in the gaseous diffusion problem of Sec. 7.5 to have a posynential problem. The original problem is

$$\min_{w} 25(0.0330 - w)(0.0071 - w)^{-1} + 0.743(1 - 2w)(0.00711 - w)^{-1}\ln(w^{-1} - 1)$$
$$+ 28.3\ln\left[(0.0071 - w)^{-1} \right]$$

*Project problem

Transform it into one in which condensation and/or partial invariance can be applied. Then solve the problem using $w = 0.009$, the solution to the approximation of Sec. 7.5, as a base case.

7-4.   P. Braga's reflected core nuclear reactor problem (Exercise 1-5) is transcendental, although it has no exponential functions. Identify the function causing this and suggest how to analyze the problem.

7-5.   Solve Exercise 1-4, Wang's power line problem, again, this time using the true equation of the caternary instead of the quadratic approximation.

7-6.   Transform Heising's bearing problem (Exercise 1-6) into posynential form (Sec. 7.3).

7-7.   Iterate the partial invariance method for the gaseous diffusion plant of Sec. 7.5, starting with $w = 0.0009$.

CHAPTER 8

Standard Sizes, Indivisible Components, and Logical Restrictions

When you have eliminated the impossible, whatever remains, however improbable, must be the truth.

Sherlock Holmes, in *The Sign of Four* (1890) by Sir Arthur Conan Doyle

Inevitably, the designer, even more than most engineers, must deal with standard sizes and other indivisible quantities, such as piece parts, processes, and projects. Verbal documents like contracts, laws, and environmental impact statements also limit the designer's choices no less than the more quantitative physical constraints and codes of the earlier chapters. This chapter develops simple methods for handling such problems, among the hardest in design.

The theory derived earlier for variables that are, in principle, infinitely divisible can generate good designs by rounding to the nearest standard size, when the variety of them is great enough. But large gaps between standard sizes may require the more careful analysis to be developed here. Since the number of rounding possibilities increases rapidly with the number of variables, effective ways are developed for ruling out entire groups of mediocre designs without computing any of them. An operations research method called "implicit enumeration" is here combined with geometric-bounding procedures to handle nonlinear design problems with discrete variables.

217

Indivisibility is a more serious matter when the designer must choose between very different component models or manufacturing methods. This leads to variables assuming only the values unity or zero according to the presence or absence of the indivisible element. To generate good designs under these circumstances, a tabular adaptation of implicit enumeration, called "lexical arithmetic," is developed. A modified version needs fewer, although more complicated, calculations.

Design requirements expressed in natural language rather than quantitative formulas, being resistant to conventional analysis, have in the past been handled informally if at all. A more scientific approach to such problems is supplied at the end of the chapter by combining lexical arithmetic with symbolic logic. The result should be useful for many logical problems other than those in design.

This chapter deals then with aspects of engineering design that have as yet seen little theoretical development. The simple, if somewhat unfamiliar, mathematics introduced here may greatly increase the designer's effectiveness, not only in finding discrete optima, but also in generating unusual but good alternative designs.

8.1. STANDARD SIZES

As noted already in Chapters 2 and 3, components in a device are often restricted to standard sizes. In the pressure vessel example of Chapter 2, rounding to a good standard design was easy because the problem was constraint bound. Neither was the pipeline example of Chapter 3 difficult, there being only two unconstrained variables. There it was necessary to examine $2^2 = 4$ cases, not an excessive number. However, generalizing the simple approach used there to N variables would involve 2^N cases, an imposition even when $N = 3$. This section, using the principle of *satis quod sufficit*, shows how to cut computation to a reasonable level without losing any good designs.

The method developed uses geometric-bounding procedures to avoid "*total* enumeration," the name given to examining all feasible candidates for the *discrete* optimum sought. Since the technique touches on all possible designs without evaluating them all in detail, it is called "*implicit* enumeration." Another more picturesque name used widely is "branch and bound."

As an example with three unconstrained variables, recall the gravel sled problem of Sec. 3.16. The cost is

$$c = 40l^{-1}w^{-1}h^{-1} + 10lw + 20lh + 40wh + 10l$$

where l, w, and h are respectively the length, width, and height of the sled. For any design, the cost is rounded to the nearest unit, and designs having the same rounded cost are considered economically indistinguishable. Lower bounds being also rounded to the nearest unit, any design indistinguishable from a lower bound is accepted as satisfactory. Thus the design $l = w = 1.28$, $h = 0.56$, having a cost of 116, is indistinguishable from a geometric lower bound computed from the following zero-degree-of-difficulty lbf which is minimum for these values.

$$c \geq 40l^{-1}w^{-1}h^{-1} + 20lw^{1/2} + 20lh + 40wh \geq 116$$

This was constructed by condensing the 2nd and 5th terms of c with equal weights: $10lw + 10l \geq 20lw^{1/2}$.

To generate a discrete variable problem, suppose that the sled is to be built from planks $\frac{1}{2}$ unit wide whose lengths must be multiples of $\frac{1}{2}$ unit. The l, w, and h can only assume the values $\frac{1}{2}$, 1, $1\frac{1}{2}$, 2, etc, making the design at hand infeasible.

Most designers would round each variable to the nearest standard size, obtaining in this case $l = w = 1.5$ and $h = 0.5$, with a cost of $c(1.5, 1.5, 0.5) = 118$. Since this is perceptably worse than the lower bound 116, a search for a better discrete design is initiated that at the same time will eliminate any designs costing 118 units or more.

To avoid missing any good possibilities, round only one variable at a time. In this case choose the height h, since its rounding interval is largest relative to its value at the continuous optimum. The standard values bracketing the continuous optimum are 0.5 and 1.0. Consider first the value requiring the lesser roundoff, namely $h = 0.5$. The other possibility ($h = 1.0$), said to be "*delayed*," will be considered later to make the search complete. The cost, depending on l and w, is

$$c(l, w, 0.5) = 80l^{-1}w^{-1} + 10lw + 20l + 20w$$

To determine how to round l and w, this single degree of difficulty function is to be bounded below by a zero degree one that is subsequently minimized. When $l = w = 1.278$, the second and third terms are 16.3 and 25.6 respectively, so the following condensation is used:

$$10lw + 20l \geq \left(\frac{10lw}{0.39}\right)^{0.39}\left(\frac{20l}{0.61}\right)^{0.61} = 29.8lw^{0.39}$$

Then

$$c(l, w, 0.5) \geq \left[(80l^{-1}w^{-1}/0.383)(29.8lw^{0.39}/0.383)(20w/0.234)^{0.61}\right]^{1/2.61}$$

$$= 116$$

indistinguishible from the continuous lower bound. Since this is less than 118, there may exist standard designs better than the one at hand. This minimum occurs at $w=0.234(116)/20=1.36$; $l=1.32$, which suggests rounding w up to its nearest standard value, namely 1.5. Of course the other possibility must be considered later to bracket this minimum.

With w rounded up, $c(l, 1.5, 0.5) = 53.3l^{-1} + 35l + 30$. This zero-degree-of-difficulty function has a minimum of 116 at $l=1.23$, which is bracketed by the standard values 1.0 and 1.5. With w and h fixed, the fact that $l=1.23$ is at the minimum implies that values less than 1.0 or more than 1.5 need not be examined, since the posynomial cost is unimodal. The case $l=1.5$ is the rounded case originally considered; the other turns out to have the same rounded cost: $c(1.0, 1.5, 0.5) = 118$.

Now begins the process of *backtracking*, which means examining the delayed alternatives previously put aside. The most recently delayed alternative having been $w=1.0$, the cost $c(l, 1.0, 0.5)$ must now be studied.

$$c(l, 1.0, 0.5) \geq 80l^{-1} + 30l + 20 \geq \left(80/\tfrac{1}{2}\right)^{1/2}\left(30/\tfrac{1}{2}\right)^{1/2} + 20 = 118.$$

Here is where some work can be saved. This lower bound of 118 means that no length exists for $w=1.0$ and $h=0.5$ that can give a design, standard or not, any better than the ones already known. Therefore the business of finding a minimizing l and examining the standard sizes bracketing it need not be done.

Finally the delayed alternative $h=1.0$ is examined.

$$c(l, w, 1.0) = 40l^{-1}w^{-1} + 10lw + 30l + 40w$$

The second and fourth terms are condensed with weights $\tfrac{1}{4}$ and $\tfrac{3}{4}$, respectively, corresponding to their relative values when $l=w=1.278$. Then

$$c(l, w, 1.0) \geq \left\{ \left(\frac{40l^{-1}w^{-1}}{\frac{4}{11}}\right) \left[\left(\frac{10lw}{\frac{1}{4}}\right)^{1/4} \left(\frac{40w}{\frac{3}{4}}\right)^{3/4} / (4/11) \right] \left(\frac{30l}{\frac{3}{11}}\right) \right\}^{4/11}$$

$$= 119$$

Since this lower bound exceeds 118, the best cost known, none of the lengths or widths for which $h=1.0$ need be examined, since all of them produce worse designs. This single bounding calculation ruled out four eventual cases, together with all the minimizations which would have generated them.

This implicit enumeration of $2^3 = 8$ possibilities only required evaluation of two cases. The other six, together with all other designs, were eliminated by the computation of two lower bounds. The original rounded design turned out to be as good as any, and the branch and bound work merely proved this. In many situations where the rounding is relatively small, the cost of the rounded design will be indistinguishable from the continuous minimum, in which case no further proof of optimality is needed.

As a matter of fact, the rounding required in this example had to be exaggerated in order to illustrate the procedure. More realistic rounding intervals for this gravel sled problem would be no more than about $\frac{1}{6}$, for which rounding would immediately give an acceptable design. This happy situation occurs often in practice, standard sizes having naturally evolved at intervals close enough to produce negligible cost change due to rounding. Otherwise, cost conscious consumers would already have demanded intermediate sizes.

Branch and bound consequently will not be needed often by designers faced with rounding problems. But the rest of the chapter deals with more severe discrete variable problems in which the designer must select from among qualitatively different components, or even choose to eliminate some of them. In such situations, the same branch and bound approach will locate the elusive discrete optimum, usually after a not unreasonable amount of effort.

8.2. REPEATED ELEMENTS: FIBONACCI SEARCH

As in the fleet design problem of Sec. 4.6, in which the number of identical ships had to be a whole number, the designer often encounters situations where some key variable is indivisible. For the structural designer, this may be the number of bays or stories in a building; for the machine designer, the number of reduction stages in a transmission; or for the process designer, the number of plates in a distillation column. Such identical components—stories, gears, or plates—will be called *repeated elements*, the general problem being to find the optimal (integer) number of repeated elements.

If the noninteger value of this optimum is easy to find by the methods already developed, then one need only round in both directions to find the best whole number. Consider, however, a case where such methods cannot be used, although the objective function is easy to optimize once the number of repeated elements is fixed. Let this objective, partially optimized wrt all variables except k, the number of repeated elements, be symbolized by $y_*(k)$. If $y_*(k)$ has but one local maximum (minimum), then

$y_\star(k)$ is said to be *unimodal* (*uninodal*). To be specific, consider the latter case, in which the objective is a cost to be minimized, and it is uninodal wrt k.

As an example, consider the desalination of seawater by multistage flash distillation. Imagine that the optimal number of flash stages, in each of which a small amount of desalted water is produced, is estimated to be around 100, give or take 10%. That is, the optimal desalination system is expected to have anywhere from 91 to 110 stages, amounting to 20 possible cases: $91, 92, \ldots, 109, 110$. It will be demonstrated now how to find the best case out of these 20 after evaluating $y_\star(k)$ only six times, provided $y_\star(k)$ is uninodal. This procedure, known in general as the *discrete Fibonacci search method*, has been proven by Kiefer to take the least number of evaluations for a given number of possibilities.

Confusion is reduced by shifting the origin so that the lowest numbered case is 1, the next 2, *etc*. A simple way to do this is to call the cases $90 + 1, \ldots, 90 + 20$, the way a surveyor marks points between stations and benchmarks.

Next use Table 8-1 to find the largest Fibonacci number less than the total number of cases. The Fibonacci numbers are easy to generate by remembering that the first two of them are 1 and 2, and each larger one is the sum of the two immediately preceding. Thus the third Fibonacci number is $1 + 2 = 3$; the fourth, $2 + 3 = 5$; the fifth, $3 + 5 = 8$, etc. In the example then, the largest Fibonacci number less than 20 is 13, the sixth in the sequence (the seventh is 21—too large). It will be seen that no more than six cost evaluations will be needed to locate the optimum number of stages out of the 20 possibilities. Evaluate the objective at this case ($90 + 13 = 103$ in the example).

Every measurement after the first will eliminate some cases from further consideration, after which the designer is left with a smaller number of candidates, one of which has already been evaluated. This is, in fact, the situation as soon as the first measurement has been taken. Thus the rule for locating the second measurement is the same as that for all later ones, the search terminating when all cases but one have been eliminated. The

Table 8-1   Location of First Fibonacci Measurement

n, maximum number of
  evaluations needed

1	2	3	4	5	6	7	8	9	10	11

F_n, location of
  first measurement

(Fibonacci numbers)	1	2	3	5	8	13	21	34	55	89	144

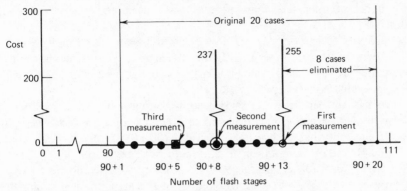

Figure 8-1 First Fibonacci elimination (20 cases)

rule is simply to measure at the largest Fibonacci number not already measured. In the example this would mean putting the second measurement at $90+8$, since $90+13$ has already been evaluated.

If the objective is uninodal, as assumed in the example, then any two evaluations will eliminate not only the worse of the two, but also all those farther away from the better one. For instance if $y_*(90+8)=237<y_*(90+13)=255$, then the optimal number of stages cannot be 13 or greater without there being at least *two* local minima, contradicting the assumption of uninodality. As Fig. 8-1 shows, this leaves 12 candidates, with $y_*(90+8)$ already known.

The general rule already stated can now be applied to the 3rd and subsequent measurements. Thus the next must be at $90+5$, the largest

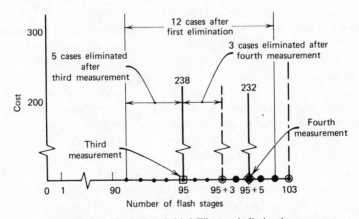

Figure 8-2 Second and third Fibonacci eliminations

Fibonacci number less than 12 that has not already been evaluated. Suppose $y_*(90 + 5) = 238 > y_*(90 + 8) = 237$, eliminating cases $90 + 1$ through $90 + 5$. The origin must now be shifted again, this time to $90 + 5 = 95$. Then, as shown in Fig. 8-2, there remain $12 - 5 = 7$ cases with an evaluation at $98 = 95 + 3$. Since the largest Fibonacci number less than 7 is $5, y_*$ is evaluated next at $95 + 5$.

Ties are not permitted in this procedure. Suppose, for instance, that the flash distillation model is considered to predict reality only to two significant figures, so that both $y_*(90 + 5)$ and $y_*(90 + 8)$ are $2.4(10^2)$, an apparent tie. Resist the temptation to eliminate not only cases $90 + 1$ through $90 + 5$, but also $90 + 8$ through $90 + 12$, although this would be correct if $y_*(90 + 5)$ and $y_*(90 + 8)$ were truly equal in the mathematical rather than the numerical sense. The example is in fact constructed to make this procedure wrong; it would eliminate the optimal case. Elimination is permitted only when the tie is broken, in the example by using the third figure so that $y_*(90 + 5) = 238 > 237 = y_*(90 + 8)$. The model may predict reality imprecisely, but comparisons between two predictions can be considered perfectly accurate for applying the elimination principle implied by the mathematical concept of uninodality. If worse came to worse, one could perturb one of the model parameters a numerically insignificant amount and recompute to break a tie. Since even pocket calculators carry 10 significant figures, true numerical ties should be rare in practice.

To continue the example, suppose $y_*(95 + 5) = 232 < 237 = y_*(95 + 3)$ so that now there remain cases $98 + 1$ through $98 + 4$, with an evaluation at $98 + 2$ as in Fig. 8-2. The largest unevaluated Fibonacci number less than 4 being 3, the fifth measurement is at $98 + 3$. If $y_*(98 + 3) = 233 > 232 = y_*(98 + 2)$, then the sixth and final measurement can only be at $98 + 1$. The optimum will be at the better of the two cases $98 + 1$ and $98 + 2$.

The sequence of numbers used here bears the nickname "Fibonacci" of Leonardo of Pisa, who first generated them back in 1202 while wondering how many pairs of offspring might be produced in a year by a pair of rabbits and their descendents. Table 8-1 shows how fast the Fibonacci numbers grow—good news for optimization as well as animal husbandry. The numbers are old, but the Fibonacci search procedure is 20th century. The modification given here is intended to be simple to remember and applicable to any number of cases. Thus the same scheme would still take exactly six measurements if there were anywhere from 13 to 20 cases, although for less than 20 the first two evaluations would not eliminate possibilities as in the full 20 case example. With 13 in fact, only one $(90 + 13$ itself) would have been eliminated, but later measurements would rapidly cut down the candidates. If there were just 21, only one more than in the example, seven measurements would be needed. In summary, if F_n is the nth Fibonacci number, generated by the recurrence relation $F_1 = F_2 =$

1; $F_n = F_{n-1} + F_{n-2}$ for $n > 2$, then m, the number of Fibonacci measurements needed to find the best out of K cases is such that $F_m \leq K < F_{m+1}$.
Designers with sporting blood can do better than this if their luck is good. Thus in the example one might risk searching the center 12 cases first (95 through 106), which in the example would have found the optimum after only five measurements. A wrong guess would be detected by convergence to one end of the interval (either 95 or 106 in the illustration), which would then require *more* search effort. Whether the possible saving merits the risk is up to the individual designer.

8.3. DISCRETE ALTERNATIVES

Now that roundoff problems caused by standard sizes have been shown to be not too difficult in practice, consider the more drastic discrete variable situation where a part may be manufactured by very different processes. Imagine for example that the designer must choose between making a part cheaply on a milling machine as opposed to grinding it to closer specifications by a more expensive process. Economy alone should not determine the choice, since the piece is only one component of an assembly whose performance may depend subtly on how well this particular component is made.

To formulate this dilemma mathematically, suppose the piece, called a "wire arm," costs 52 units if ground and only 42 units if milled. Then the cost can be written $52g + 42m$, where g is understood to be unity if the grinder is used and zero if it is not. Similarly $m = 1$ if the part is milled, and $m = 0$ if not. Such variables that can only assume the values zero or unity are said to be *bivalent*. The fact that one, and only one, of the two processes can be used is expressed by the simple equation $g + m = 1$, an example of a *logical* constraint. To complicate things further, suppose the part need not be finished by either process. Then the cost function would not change, but the logical constraint would become $g + m \leq 1$, since there is now the possibility of using neither process. As the problem is now formulated, the least-cost solution would obviously be $g = m = 0$.

More realistically, there may be a constraint that the semitolerance for the part not exceed 5 length units (e.g., millimeters or thousandths of an inch). If the unfinished part fluctuates ± 10 units, a milled part ± 4 units, and a ground part ± 2 units, then introducing the bivalent variable u, unity for an unfinished part but zero otherwise, permits writing the semitolerance constraint as $2g + 4m + 10u \leq 5$. Clearly $u = 1$ would not be feasible, so $u_* = 0$ and the problem remaining is to minimize the cost subject to the tolerance constraint $2g + 4m \leq 5$ and the logical constraint $g + m = 1$. But the logical constraint is a strict equality, so it can be used to eliminate a variable, say g, leaving the objective as $52 - 10m$ constrained only by $2m \leq 5 - 2 = 3$. Since the cost decreases monotonically with m, this variable must be made as large as possible, namely unity, which satisfies the tolerance constraint. The solution is then to mill the part, since this is the cheapest process satisfying the tolerance requirement.

The preceding example intended merely to show how to formulate discrete choice problems in terms of bivalent variables and then to perform elementary simplifications. Most designers could have guessed the answer without any analysis in this simple case, but consider now a more realistic tolerance assignment problem due to Smathers and Ostwald ("Optimization of Component Functional Dimensions and Tolerances," ASME paper 72-DE-18) which has eight parts, 20 manufacturing processes, and two tolerance loops, giving in all 1296 possible assignments of parts to processes. Here, as in most discrete decision problems, preliminary simplifications are very powerful, cutting out 1272 possibilities, leaving only 24 for more sophisticated analysis.

The problem of optimizing a small design will be demonstrated in the allocation of tolerances to an eight-component winder assembly (Fig. 8-3). Selection of tolerances is to achieve least-cost manufacture with choice of the processing based upon available equipment. For each length tolerance, two or three processes are shown.

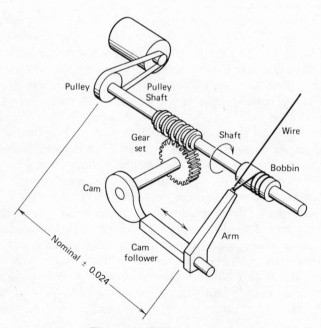

Figure 8-3 Winder assembly

In the usual situation, many process-tolerance-cost choices are available. This information must be provided before the tolerances can be assigned and is given along with the process designations, in Table 8-2. The design specifies a maximum material condition tolerance of ± 0.024 in. (0.048 in. overall) for the lateral direction. Other axes may be handled in an identical manner but are not shown for the sample problem.

Table 8-2 Component Process Designation, Tolerance, and Cost

Component	Component Identification	Process Description	Process Designation		Half Natural Tolerance, in.	Process Tolerance Cost
Pulley	1	Lathe	A	10	0.002	$ 0.63
		Turret Lathe	C	11	0.006	0.43
Pulley shaft	2	Turret Lathe	D	20	0.006	0.42
		Cut Off	E	21	0.008	0.24
Gear set	3	Gear Generator	F	30	0.001	1.05
		Gear Generator	G	31	0.002	0.91
		Univ. Milling Machine	H	32	0.004	0.65
Shaft	4	Lathe	I	40	0.008	0.22
		Turret Lathe	J	41	0.010	0.20
		Cut Off	K	42	0.016	0.12
Bobbin	5	Lathe	L	50	0.002	0.69
		Turret Lathe	M	51	0.006	0.21
		Chucker	O	52	0.008	0.10
Cam	6	Cam Grinder	P	60	0.006	1.41
		Cam Cutter	Q	61	0.008	1.00
Cam follower	7	Lathe	R	70	0.002	0.61
		Turret Lathe	S	71	0.006	0.23
		Cut Off	T	72	0.008	0.10
Wire arm	8	Grind	U	80	0.002	0.52
		Mill	V	81	0.004	0.42

This problem will now be formulated generally in terms of bivalent variables and then written specifically with the numbers of the example. Let there be P parts indexed by $p = 1, 2, \ldots, P$, and let the pth part be manufacturable by exactly one of $1 + Q(p)$ processes, indexed by $q = 0, 1, \ldots, Q(p)$. Let t_{pq} be the semitolerance and c_{pq} the cost if part p is manufactured by process q. Arrange the indices so that $t_{p0} < t_{p1} < \cdots < t_{pQ(p)}$, and $c_p > c_{p1} > \cdots > c_{pQ(p)}$. Assume that x_{pq} is a variable taking on the value unity (zero) if part p is (is not) made by process q. Then the total manufacturing cost to be minimized is

$$c \equiv \sum_{p=1}^{P} \sum_{q=0}^{Q(p)} c_{pq} x_{pq} \tag{0'}$$

In the example, $P = 8$, and

$$c = (63x_{10} + 43x_{11}) + (42x_{20} + 24x_{21}) + (105x_{30} + 91x_{31} + 65x_{32})$$
$$+ (22x_{40} + 20x_{41} + 12x_{42}) + (69x_{50} + 21x_{51} + 10x_{52}) + (141x_{60} + 100x_{61})$$
$$+ (61x_{70} + 23x_{71} + 10x_{72}) + (52x_{80} + 42x_{81})$$

There are R constraints, indexed $r = 1, \ldots, R$, reflecting known upper bounds u_r on assembly tolerances. Let $P(r)$ be the subset of indices for the parts involved in constraint r. The the *tolerance constraints* are

$$f_r \equiv \sum_{p \in P(r)} \sum_{q=0}^{Q(p)} t_{pq} x_{pq} \leq u_r \qquad r = 1, \ldots, R \tag{r'}$$

In the example, $R = 2$, $P(1) = \{1, 2, 3, 5, 6, 7, 8\}$, $P(2) = \{1, 2, 3, 4, 5\}$, and the constraints are

$$f_1 \equiv (2x_{10} + 6x_{11}) + (6x_{20} + 8x_{21}) + (x_{30} + 2x_{31} + 4x_{32})$$
$$+ (2x_{50} + 6x_{51} + 8x_{52}) + (6x_{60} + 8x_{61}) + (2x_{70} + 6x_{71} + 8x_{72})$$
$$+ (2x_{80} + 4x_{81}) \leq 24 \tag{1'}$$

$$f_2 \equiv (2x_{10} + 6x_{11}) + (6x_{20} + 8x_{21}) + (x_{30} + 2x_{31} + 4x_{32})$$
$$+ (8x_{40} + 10x_{41} + 16x_{42}) + (2x_{50} + 6x_{51} + 8x_{52}) \leq 24 \tag{2'}$$

Since each part must be made by exactly one process, there are P additional equality constraints.

$$f_{R+p} \equiv \sum_{q=0}^{Q(p)} x_{pq} = 1 \qquad p = 1, \ldots, P \tag{R'+p'}$$

In the example, these are:

$$x_{10} + x_{11} = 1 \tag{3'}$$

$$x_{20} + x_{21} = 1 \tag{4'}$$

$$x_{30} + x_{31} + x_{32} = 1 \tag{5'}$$

$$x_{40} + x_{41} + x_{42} = 1 \tag{6'}$$

$$x_{50} + x_{51} + x_{52} = 1 \tag{7'}$$

$$x_{60} + x_{61} = 1 \tag{8'}$$

$$x_{70} + x_{71} + x_{72} = 1 \tag{9'}$$

$$x_{80} + x_{81} = 1 \tag{10'}$$

8.4. EXCESS TOLERANCE TRANSFORMATION

The first simplification comes from transforming to *excess tolerances* e_{pq} and *cost savings* s_{pq} defined by

$$e_{pq} \equiv t_{pq} - t_{p0} > 0$$

and

$$s_{pq} \equiv c_{pq} - c_{p0} > 0 \qquad q = 1, \ldots, Q(p)$$

Then elimination of all P variables x_{p0}, using Eqs. $(R' + p')$, gives the equivalent problem of maximizing the cost saving s

$$\sum_p c_{p0} - c \equiv s \equiv \sum_{p=1}^{P} \sum_{q=1}^{Q(p)} s_{pq} x_{pq} \geq 0 \tag{0}$$

subject to excess tolerance constraints

$$g_r \equiv \sum_{p \in P(r)} \sum_{q=1}^{Q(p)} e_{pq} x_{pr} \leq e_r \equiv u_r - \sum_{p \in P(r)} t_{p0} \tag{r}$$

as well as logical constraints preventing assigning the same part to more than one process. These replace the equalities $(R' + p')$.

$$\sum_{q=1}^{Q(p)} x_{pq} \geq 1 \tag{$R + p$}$$

Any of these having $Q(p)=1$ are, of course, trivial and may be deleted.

In the example, these functions have 12 rather than the original 20 variables.

$$s = 20x_{11} + 18x_{21} + 16x_{31} + 40x_{32} + 2x_{41} + 10x_{42}$$
$$+ 48x_{51} + 59x_{52} + 41x_{61} + 38x_{71} + 51x_{72} + 10x_{81} \tag{0}$$
$$g_1 = 4x_{11} + 2x_{21} + x_{31} + 3x_{32} + 4x_{51} + 6x_{52} + 2x_{61}$$
$$+ 4x_{71} + 6x_{72} + 2x_{81} \le 24 - (2+6+1+6+2+2+2) = 3 \tag{1}$$
$$g_2 = 4x_{11} + 2x_{21} + x_{31} + 3x_{32} + 2x_{41} + 8x_{42}$$
$$+ 4x_{51} + 6x_{52} \le 24 - (2+6+1+8+2) = 5 \tag{2}$$

There are four nontrivial logical constraints

$$x_{31} + x_{32} \le 1 \qquad x_{41} + x_{42} \le 1 \qquad x_{51} + x_{52} \le 1 \qquad x_{71} + x_{72} \le 1$$

To preserve feasibility, any variable x_{pq} whose excess tolerance exceeds the total e_r for any constraint must be set to zero.

$$e_{pq} > e_r \text{ implies } x_{pq} = 0 \tag{11}$$

In the example this immediately eliminates six variables

$$x_{11} = x_{42} = x_{51} = x_{52} = x_{71} = x_{72} = 0 \tag{4}$$

Moreover, this exhausts the alternatives for parts 1, 5, and 7, so that

$$x_{10} = x_{50} = x_{70} = 1$$

The only nontrivial logical constraint remaining is

$$x_{31} + x_{32} \le 1 \tag{5}$$

Substitution of Eq. (4) into Eqs. (0), (1), and (2) gives a problem in only six variables.

$$s = 18x_{21} + 16x_{31} + 40x_{32} + 2x_{41} + 41x_{61} + 10x_{81} \tag{6}$$
$$g_1 = 2x_{21} + x_{31} + 3x_{32} + 2x_{61} + 2x_{81} \le 3 \tag{7}$$
$$g_2 = 2x_{21} + x_{31} + 3x_{32} + 2x_{41} \le 5 \tag{8}$$

Thus it is easy to find the allocation consuming the least total tolerance, albeit at maximum cost, by merely making each part to the tightest tolerance in the most expensive way. A convenient simplification then is to work with the excess tolerance for each part, obtained by subtracting the minimum tolerance from that for all other processes. These excess tolerances then must sum to no more than the assembly tolerance minus the least total tolerance. This transformation eliminates as many variables as there are parts, always an important simplification. Moreover, it makes the objective that of maximizing the cost saving compared to the maximum cost (minimum tolerance) case.

With the problem in excess tolerance form, one can easily see if any manufacturing processes require more excess tolerance than is available. Such processes, unfortunately those saving the most money, can be eliminated as infeasible. Often this in turn will rule out all but the most precise process for a particular part, simplifying the problem further.

8.5. CONSTRAINT REDUNDANCE

A second level of simplification involves the constraints, both those describing tolerance allocation and those specifying that each part be made by exactly one process. One often may find constraints that are redundant, i.e., that are automatically satisfied whenever other constraints are, and consequently delete them. Detailed verbalizations of rigorous ways of detecting redundance are given later. Since a part restricted only by deleted constraints becomes unconstrained, it may be manufactured in the least expensive way, a further simplification.

In the example, the first three parts appear in both constraints g_1 and g_2 (Eqs. 7 & 8), while the latter has more excess tolerance available. The following theorem proves that g_2 (Eq. 8) is redundant.

Theorem 1 (Tolerance Constraint Redundance)

Suppose

$$g_1(x) = h_1(x) + h_3(x) \le e_1,$$

$$g_2(x) = h_1(x) + h_2(x) \le e_2, \quad \text{and}$$

$$h_2(x) \le e_2 - e_1,$$

where $h_1(x)$, $h_2(x)$, and $h_3(x)$ are all nonnegative functions of the bivalent variables x. Then any solution $\bar{x}$ satisfying $g_1(\bar{x}) \le e_1$ also satisfies $g_2(\bar{x}) \le$

Proof. Since $h_3(\bar{x}) \geq 0$, $h_1(\bar{x}) \leq e_1$. Hence $g_2(\bar{x}) \leq e_1 + (e_2 - e_1) = e_2$.

In the example, let $h_1 \equiv 2x_{21} + x_{31} + 3x_{32}$, $h_2 \equiv 2x_{41} \leq 2 = e_2 - e_1$ to prove that g_2 is redundant. Moreover, logical constraint (5) can also be proven redundant by noticing that it cannot be violated unless g_1 is also violated.

Theorem 2 (Logical Constraint Redundance)

If $e_{ps} + e_{pt} > e_r$ for some part p and every pair s and t of manufacturing processes for p, then the logical constraint $\sum_{q=1}^{Q(r)} x_{pq} \leq 1$ is redundant.

Proof. If x' satisfies $g_r \leq e_r$, then for every pair s and t, $e_{ps}x'_{ps} + e_{pt}x'_{pt} \leq e_r$, which implies that no two variables x'_{ps} and x'_{pt} can be equal to unity simultaneously. Hence x' satisfies the logical constraint automatically.

Theorem 2 permits deletion of constraint (5). Whenever a constraint is deleted, one should examine the variables in the objective function to see if any are now unconstrained. Because all objective function coefficients are positive, money can be saved by setting such variables equal to unity. In the example the deletion of redundant constraints (5) and (8) leaves x_{41} unconstrained, so set

$$x_{41} = 1$$

This takes care of part number 4 and reduces the problem to the following one in five variables.

$$\max s = 18x_{21} + 16x_{31} + 40x_{32} + 41x_{61} + 10x_{81} + 2$$

subject to

$$g_1 = 2x_{21} + x_{31} + 3x_{32} + 2x_{61} + 2x_{81} \leq 3$$

For future reference, this principle is stated as a theorem with the proof omitted.

Theorem 3 (Unconstrained Variables)

Any variable not appearing in a constraint should be set to unity.

8.6.   FURTHER ELIMINATION

Although the single constraint remaining in principle permits the constraint function g_1 to be strictly less than three, no such solution can be optimal. For if $g_1 = 2$, then a better solution is obtained by setting $x_{31} = 1$, making $g_1 = 3$, and similarly if $g_1 = 1$, a better solution is achieved by setting to unity any variable with constraint coefficient 2. Hence no generality is lost in replacing the inequality by a strict equality and solving for x_{31}, which has a unit coefficient. Then x_{31} can be eliminated from the objective to give

$$s = 2 + 18x_{21} + 16[3 - 2x_{21} - 3x_{32} - 2x_{61} - 2x_{81}] + 40x_{32} + 41x_{61} + 10x_{81}$$
$$= 50 - 14x_{21} - 8x_{32} + 9x_{61} - 22x_{81}$$

This expression can be converted to one with only positive coefficients by introducing the logical complements of those variables with negative signs (Hammer and Rudeanu). In this case, substitute the bivalent variables

$$x_{21} \equiv 1 - \bar{x}_{21} \qquad x_{32} \equiv 1 - \bar{x}_{32} \qquad x_{81} \equiv 1 - \bar{x}_{81}$$

to obtain

$$s = 6 + 22\bar{x}_{81} + 14\bar{x}_{21} + 9x_{61} + 8\bar{x}_{32}$$

Notice that the variables have been rearranged in descending order of coefficients (Hammer and Rudeanu). Thus one can save the most (22¢) by *not* using process x_{81} ($\bar{x}_{81} = 1$ means $x_{81} = 0$ and $x_{80} = 1$) whereas *using* x_{61} saves 9¢.

A constraint remains, however. The variable x_{31} eliminated must still be bivalent, which gives, in terms of the original variables,

$$1 \geq 3 - 2x_{21} - 3x_{32} - 2x_{61} - 2x_{81}(= x_{31}) \geq 0$$

From this is obtained the following double-ended inequality, rewritten in the new variables of the objective function.

$$4 \leq 2\bar{x}_{81} + 2\bar{x}_{21} - 2x_{61} + 3\bar{x}_{32} \leq 5$$

The problem is now to maximize the cost saving subject to this double inequality.

One advantage of this new formulation is that it has only four variables instead of five. On the other hand, it also has two inequalities instead of one, and the new constraints contain a negative sign, all of which at first glance may seem less desirable. However, the variable portion is identical for both constraints, and the range of the feasible region is only a unit interval.

Most important, the change of variables now makes the optimal solution apparent. The maximal unconstrained solution ($\bar{x}_{81} = \bar{x}_{21} = x_{61} = \bar{x}_{32} = 1$) is easily verified as satisfying the double-ended constraint, so it is also the constrained maximum. Substitution of these values gives $x_{31} = 1$; the maximum cost saving is $6 + 22 + 14 + 9 + 8 = 59\cent$.

In terms of the original lettered variables, the optimal production plan verifies that found by Smathers and Ostwald, namely, ADGJLQRU. All but three parts are made by the highest precision process available; the exceptions are the gear set, made on the gear generator put to its highest tolerance; the shaft, made on the turret lathe; and the cam, made on the cam cutter instead of the grinder. Equivalently, the optimal semitolerances of the eight parts, which is how the designer specifies the manufacturing processes, are 2, 6, 2, 10, 2, 8, 2, and 2 units, respectively.

The design remains optimal over a range of process cost savings. Four parts (1, 4, 5, and 7) will be held to minimum tolerance no matter how the costs fluctuate. The cost-saving coefficients show how much the other four cost savings can change before the design becomes nonoptimal. Thus the term $22\bar{x}_{81}$ implies that, other costs unchanged, the cost saving for part 8 would have to increase 22¢ (from 10¢ to 32¢) before it would be more economical to mill (x_{81}) rather than grind (x_{80}) the wire cam. On the other hand, the term $9x_{61}$ says that the cam should be cut rather than ground as long as the cost saving does not decrease more than 9¢ (from 41¢ to 32¢). All of this presumes that the cost coefficient of x_{31}, the variable eliminated, is constant. However, the optimal design is relatively sensitive to this coefficient, since it is multiplied by the corresponding excess tolerance. Thus if all the cost savings except that of x_{31} stayed constant, the coefficient s_{31} of the latter must remain in the range

$$13\tfrac{1}{3} = \max\left\{ \tfrac{10}{2}, \tfrac{18}{2}, \tfrac{40}{3} \right\} \leq s_{31} \leq \tfrac{41}{2} = 20.5$$

which is only 7¢ in width. This sensitivity analysis gives an idea of how likely the design will remain optimal in the face of uncertainty in the economic data.

This study has shown how a little analysis can simplify a tolerance assignment; for an assembly of eight parts, two tolerance loops, and 20 manufacturing processes, the best design is found and proven optimal without ever evaluating the cost. This is an improvement over dynamic programming, the best competing discrete optimization procedure, which requires computation equivalent to 130 function evaluations for the same problem.

Although obtaining a complete solution is in this case fortuitous, the same principles can simplify other tolerance problems, usually reducing the computation needed. One first eliminates a process for each part by formulating the problem in terms of excess tolerances and cost savings. This step often determines some of the tolerances immediately. Then, if possible, redundant constraints are identified and deleted, which occasionally leaves tolerances unrestricted, permitting their ·fixation at optimal values. In this example a third step, namely solution of the single constraint remaining, permitted immediate identification of the optimum. More generally, the first step can always be carried out; the second, sometimes; whereas the third only rarely suffices to find the answer.

Since the objective and all constraints are linear, the problem could have been solved by linear programming, provided inequality constraints $x_{pq} \leq 1$ were included. Remarkably, the solution would have been correct in this case, for the minimum of the continuous problem happens to have values of zero or unity for every variable. This is a coincidence, however, and would not have occurred if the cost saving for x_{31} were reduced from 16 to 11. Then the simplex algorithm would find it economical to replace x_{31} with x_{32}, which would want to take the fractional value $\frac{1}{3}$, for an infeasible saving of $56\frac{1}{3}$, compared to 55 for the feasible maximum. Linear programming is therefore not a reliable way to solve bivalent problems. But if a linear programming code is handy, a quick linear programming check won't hurt, may be lucky, and will provide a bound on the solution. For preliminary estimates, this may be all that's needed.

The rest of the chapter deals with how to finish the job when these procedures do not lead directly to the answer. Implicit enumeration is needed for this more general situation.

8.7.   LEXICAL ARITHMETIC

Although most designers would not mind using the simplification scheme of Sec. 8.4, some might balk at having to detect and prove constraint redundance or activity as in Secs. 8.5 and 8.6. To take care of this valid objection, the rest of the chapter develops and illustrates a tabular scheme, based on implicit enumeration, for solving bivalent problems. The method is readily extended, with slight modification, to interesting logical expressions derived from natural language. Mastery of the procedure gives the engineer a simple rigorous way to analyze the verbal, legal, contractual, and even psychological aspects of design.

The name of the method is *lexical arithmetic*. It is a collection of rules and shortcuts resembling those of such algorithms of arithmetic as long division taught in elementary school. This tabular procedure, intended to be useful in the "back of the envelope" situations that the engineer often faces in practice, is somewhat simpler than keeping an official baseball scorebook, although it does require reasonable care and some addition and subtraction. It follows the ideas of Hammer and Rudeanu, adapting them to the needs of pencil-and-paper designers.

The method is called "lexical" because the terms of the objective function are arranged in order of their individual values, much as words in a dictionary, or *lexicon*. In attempting to assign values to the bivalent variables, one grants priority to the more valuable terms, meanwhile checking the constraints for feasibility. The procedure is effective because often a given partial assignment dictates the values of the rest of the variables needed to preserve feasibility. Also, the lexical ordering of the solutions often makes clear when further feasible assignments cannot possibly generate a solution any better than the one already in hand. Lexical arithmetic is a simple framework for doing implicit enumeration.

The lexical approach requires each term of the objective, which throughout this chapter will be *max*imized, to be positive. Every problem can be put into this form, for if a term Ax has the coefficient A negative, then substituting the identity $x \equiv 1 - \bar{x}$, where the bivalent variable $\bar{x}$ is called the *logical complement* of x, gives $Ax = A - A\bar{x} = A + |A|\bar{x}$. Since $\max(Ax) = \max(A + |A|\bar{x}) = A + \max(|A|\bar{x})$, one can solve the original problem by maximizing $|A|\bar{x}$.

Once each term has a positive coefficient, the variables are rearranged so that the term with the highest coefficient is first (i.e., on the left), etc. Then the *lexical table* is constructed with a column for each term, as in Table 8-3. Immediately below the objective are the coefficients of the inequality constraints. Notice that any strict equalities in the original problem have been eliminated in the preliminary simplification of Sec. 8.3. To the right of the working columns are spaces to be explained later. Roughly speaking, the left one records which terms are nonzero in a given trial solution, while

Table 8-3   Lexical Table Initial Layout

Index		Original	61	32	21	31	81	41			
		Ordered	1	2	3	4	5	6			
Cost saving			41	40	18	14	10	2			
	Arm		2	3	2	1	2		≤ 3		
Constraints	Shaft			3	2	1		2	≤ 5		
	Gear			1		1			≤ 1	Solution	Dominated by

the right one tells of other solutions that may "dominate" the one on this line by having a better value of the objective. Table 8-3 shows the initial layout of the lexical table for the tolerance assignment problem already solved.

Consider the consequences of setting x_1 (formerly x_{61}), the variable with the largest objective coefficient, to unity: $x_1 = 1$. The first constraint now reads

$$2x_1 + 3x_2 + 2x_3 + x_4 + 2x_5 = 2 + 3x_2 + 2x_3 + x_4 + 2x_5 \leq 3$$

Subtraction of the coefficient 2 of x_1 in this constraint gives the new inequality $3x_2 + 2x_3 + x_4 + 2x_5 \leq 1$. Clearly then, $x_2 = x_3 = x_5 = 0$. All of these results are recorded as ones and zeroes on the top empty line of the lexical table, indicated here to save space as

x_1	x_2	x_3	x_4	x_5	x_6
1	0	0		0	

Constraints 2 and 3 are checked in the same manner, but no new conclusions follow.

It is now possible to assign ones to some of the blank spaces, which represent variables whose values have not yet been assigned. The rule is to assign a one to the first, i.e., leftmost, blank, since this corresponds to the unassigned variable with the largest, and therefore the most desireable, objective coefficient. Here this is equivalent to setting $x_4 = 1$ temporarily, consideration of the alternative $x_4 = 0$ being delayed until later. Checking the constraints shows that no other variable (only x_6 remains) is forced to zero by this assignment. Thus a 1 can be entered in column x_4 of the lexical table, on the second line to avoid confusing it with the first assignment ($x_1 = 1$).

x_1	x_2	x_3	x_4	x_5	x_6
1	0	0		0	
			1		

Repetition of the rule assigning a 1 to the next open blank gives $x_6 = 1$, so that now a complete feasible solution has been generated. In it, $x_1 = x_4 = x_6 = 1$, and $x_2 = x_3 = x_5 = 0$, which is indicated in the solution column by 1,4,6, the indices of the variables given unit values.

x_1	x_2	x_3	x_4	x_5	x_6	solution
1	0	0		0		
			1			
					1	1,4,6

A solution having been generated, it is now time to backtrack. Each of the ones in the feasible solution $(1,4,6)$ was assigned temporarily, there having been no advance assurance that the optimum solution would have ones in those particular places. It still remains to examine the delayed alternatives in which these variables are zero rather than one. A systematic way is to reset to zero the variable most recently set to one—x_6 in the example. This always corresponds to the 1 farthest to the right in the lexical table. The backtracking is indicated by placing a 0 immediately under the rightmost 1 as follows:

x_1	x_2	x_3	x_4	x_5	x_6	solution	dominated by
1	0	0		0			
			1				
						1	1,4,6	
						0	1,4	1,4,6

Notice that a new feasible solution $(1,4)$ has been generated by this first backtrack. However, this new solution cannot be as good as the solution $(1,4,6)$ obtained earlier, since $y(1,4,6)=y(1,4)+A_6>y(1,4)$. This fact is indicated by writing $(1,4,6)$ in the "dominated by" column on the same line as the inferior (or *dominated*) solution $(1,4)$.

Of course, any solution generated by backtracking on a 1 in the last (rightmost) available space is always dominated by the solution preceding, and so really need not be recorded. Instead, such a 1 will always be entered on the same line as the 1 preceding it, so that generation of the first feasible solution is indicated in only two lines as follows:

x_1	x_2	x_3	x_4	x_5	x_6	solution
1	0	0		0		
			1		1	1,4,6

Since the next 1 to the left is in column x_4, backtracking would now involve setting x_4 to zero. But assigning zeroes in place of any of the existing 1s in x_4 and x_6 could never produce a feasible solution better than $(1,4,6)$, so there is no need to write down the dominated solutions. The rule then is to locate the rightmost 1 having 0s on the same line (x_1 in the example) and backtrack there by setting that variable to zero. The 0 is written on a new line in the same column. Then, as before, a 1 is assigned on that line in the leftmost blank space, making $x_2=1$. This immediately makes the first constraint tight, forcing zeros for x_3, x_4, and x_5. Since this leaves only one blank (at x_6), a 1 can be entered there immediately, giving the feasible solution $(2,6)$. But since $A_1>A_2$ and $A_4>A_6$, this new solution cannot be as good as $(1,4,6)$. Formally,

$$y(1,4,6) = A_1 + A_4 + A_6 > A_2 + A_6 = y(2,6)$$

Hence the new solution is dominated as shown.

x_1	x_2	x_3	x_4	x_5	x_6	solution	dominated by
1	0	0		0			
			1		1	1,4,6	
0	1	0	0	0	1	2,6	1,4,6

The backtrack rule now seeks the 1 closest to x_6 on the left which has 0s in the same row. This is in column x_2, and so a 0 is put under the 1 in column x_2 to begin a new line. The 0 on line 3 of the x_1 column is understood to hold for all lines below it, since both possibilities for x_1 have now been exhausted. The next blank available for a 1 is in the x_3 column, and this forces a 0 for x_5. Although this leaves two blanks remaining, they don't have to be filled in because even if they were both unity, they could not generate a solution as good as $(1,4,6)$. Consequently, further checking for feasibility is unnecessary, and the spaces are filled with xs for the record.

x_1	x_2	x_3	x_4	x_5	x_6	solution	dominated by
1	0	0		0			
			1		1	1,4,6	
0	1	0	0	0	1	2,6	1,4,6
	0	1	x	0	x	3,4,6	1,4,6

Backtracking now on x_3 leaves only three blanks. Since they are at x_4, x_5, and x_6, no new solution can possibly be better than $(1,4,6)$, so the blanks can be crossed out. There being no more 1s for backtracking, the search is complete. The solution $(1,4,6)$ is maximal, *independently of the numerical values of the cost-saving coefficients*. The same solution maximizes the cost saving, even if the coefficients change, provided their order does not. The full lexical table follows.

x_1	x_2	x_3	x_4	x_5	x_6	solution	dominated by
1	0	0		0			
			1		1	1,4,6	
0	1	0	0	0	1	2,6	1,4,6
	0	1	x	0	x	3,4,6	1,4,6
		0	x	x	x	4,5,6	1,4,6

The activity of the first constraint and the redundancy of the second could have been detected after the fact by noticing that no constraint but the former is ever tight. But all the failure to detect the redundance in advance did was make necessary the checking of the redundant constraint each time.

8.8. DISCRETE LOCAL OPTIMA

The concept of a local maximum can be defined for problems in bivalent variables. A *local maximum* is a point where changing any *single* component (i.e., a 0 to a 1 or vice versa) gives a new point that is either infeasible or has its objective value less than at the original point. Lexical arithmetic generates local maxima. In the preceding example, one local maximum dominated all others, and so it was found to be the global maximum without any need to evaluate the objective. This rare situation arose because the constraints permitted but few feasible solutions, and so the computations were mercifully short.

Consider now a less tightly constrained situation. Suppose the previous problem is modified by increasing the total semitolerance from 24 to 27 units. After simplification the new problem has all 12 variables in it, and none of the constraints are redundant, giving the lexical Table 8-4.

One new situation occurs, on the line marked by the arrow in Table 8-4. There, when backtracking assigns $x_1 = x_2 = x_3 = 0$ and $x_4 = 1$, the remaining variables are still unrestricted, and so no zeroes are generated on the line. In such cases, which arise often when the constraints permit many feasible solutions, there is no need to start a new line for the next 1 to be assigned. Instead, the 1 is entered on the same line, and then the remaining variables are checked to see if any are driven to zero. In this case, $x_4 = x_5 = 1$ implies that all variables x_6-x_{11} must be zero. The final variable is not assigned, since $(4, 5, 12)$ would be dominated by $(3, 4, 12)$ as noted on the right. Notice that the backtracking on the next line is at the rightmost 1 assigned, namely at x_5, not x_4. The backtracking continues with x_4 held at 1 as long as possible local maxima can be generated, and then finally, nine lines below, x_4 is reset to zero.

Lexical arithmetic generated five local maxima not directly comparable to each other, so the objective function must be evaluated for each of them. Thus $y(1) = 59$; $y(2, 11) = 61$; $y(3, 4, 12) = 91$; $y(4, 6, 11) = 89$; and $y(4, 8, 9, 12) = 75$. The global maximum is therefore $(3, 4, 12)$ where the cost saving is 91 units per assembly.

Unlike the first example, this solution is sensitive to at least some of the cost coefficients, even when the order of the savings coefficients does not change. Thus $y(3, 4, 12) > y(4, 6, 11)$ only as long as $A_3 + A_{12} > A_6 + A_{11}$, and the current difference is only two units.

Table 8-4   Lexical Table for 27 Units Total Tolerance

Index		1	2	3	4	5	6	7	8	9	10	11	12			
Index	Original	52	72	51	61	32	71	11	21	31	81	42	41			
	Ordered	1	2	3	4	5	6	7	8	9	10	11	12			
Cost saving		59	51	48	41	40	38	20	18	14	10	10	2			
Con-straints	Arm	6	6	4	2	3	4	4	2	1	2			≤ 6		
	Shaft	6		4		3		4	2	1		8	2	≤ 8		
	Gear				1				1					≤ 1	Solution	Dominated by
		1	0	0	0	0	0	0	0	0	0	0	1		1, 12	
		0	1	0	0	0	0	0	0	0	0	1	0		2, 11	
			0	1	1	0	0	0	0	0	0	0	1		3, 4, 12	
					0				1	0	0		X		3, 8, 12	3, 4, 12
									0	1	0		X		3, 9, 12	3, 4, 12
										0	X		X		3, 10, 12	3, 4, 12
→				0	1	1	0	0	0	0	0	0	X		4, 5, 12	3, 4, 12
						0	1	0	0	0	0	1	0		4, 6, 11	
							0	1	0	0	0	0	X		4, 7, 12	4, 6, 11
								0	1	1	0		X		4, 8, 9, 12	
										0	X		X		4, 8, 10, 12	4, 8, 9, 12
									0	1	X	0	X		4, 9, 10, 12	4, 8, 9, 12
										0	X	X	X		4, 10, 11, 12	4, 8, 9, 12
					0	1	0	0	X	0	X	0	X		5, 8, 10, 12	4, 8, 9, 12
						0	1	0	1	0	0	0	X		6, 8, 12	4, 8, 9, 12
									0	1	0	X	X		6, 9, 11, 12	4, 8, 9, 12
										0	X	X	X		6, 10, 11, 12	4, 8, 9, 12
							0	1	1	0	X	0	X		7, 8, 10, 12	4, 8, 9, 12
									0	X	X		X		7, 9, 10, 12	4, 8, 9, 12
								0	1	X	X	0	X		8, 9, 10, 12	4, 8, 9, 12
									0	X	X	X	X		9, 10, 11, 12	4, 8, 9, 12

241

8.9.   OBJECTIVE FUNCTION BOUNDING

When, as in the examples, the objective function is easy to evaluate, it can be used to construct lower bounds on the maximum to rule out groups of solutions. Thus in the last example, the cost saving of 91 for $(3, 4, 12)$ is a lower bound on the maximum, which may have a higher value but certainly not a lower one. Then as newer solutions are generated, the actual saving as well as upper bounds on the potential savings can be computed. Whenever such an upper bound falls below the known lower bound, computation of that solution can be ended as unprofitable.

For instance, immediately after $(3, 4, 12)$ has been generated, backtracking holds $x_3 = 1$, making $x_5 = x_6 = x_7 = x_{11} = 0$, but x_4 is reset to zero. The unassigned variables are x_8, x_9, x_{10}, and x_{12}, and the potential cost saving if

Table 8-5   Numerical Branch and Bound

Index	Original	52	72	51	61	32	71	11	21	31	81	42	41	≤ 6
	Ordered	1	2	3	4	5	6	7	8	9	10	11	12	
Cost saving		59	51	48	41	40	38	20	18	14	10	10	2	
Con-straints	Arm	6	6	4	2	3	4	4	2	1	2			≤ 6
	Shaft	6		4		3		4	2	1		8	2	≤ 8
	Gear				1					1				≤ 1 Dominated by
		61	0	0	0	0	0	0	0	0	0	0	61	91
		0	63	0	0	0	0	0	0	0	0			
												61	0	91
			0	133		0	0	0				0		
				91				0	0	0			91	
			92	0				76	0					91
			0	156	56	0	0	0		0	0			91
				116	0	52	0	0	0	0				91
				78	0									91
					0	70	0	0		0		0		91
					0	92	0							
						78	0							91
							0	78						91

all of them could be set at 1 is $A_8 + A_9 + A_{10} + A_{12} = 18 + 14 + 10 + 2 = 44$. Since $x_3 = 1$, there is already a saving of 48, giving an upper bound of $48 + 44 = 92$ on further solutions generated with $x_1 = x_2 = 0$, $x_3 = 1$, and $x_4 = 0$. Since this upper bound exceeds 91 and the lower bound is already in hand, the computation can proceed. The first unassigned variable x_8 is therefore set to 1, which causes x_9 to be zero. Hence the value of $A_9(= 14)$ must be deducted from the upper bound, which becomes $92 - 14 = 78$. This is less than 91, so there is no need to complete the solution. Backtracking proceeds by setting x_8 to zero, but the upper bound for this partial solution is $92 - 18 < 91$, so this line can also be terminated, permitting backtracking all the way to x_3.

The computations can be tabulated on the same form as for lexical arithmetic. Whenever a variable is set to 1, the upper bound for the partial solution is entered. Any time a feasible solution is completed, the rightmost number becomes a lower bound on the maximum. If an upper bound falls below the current lower bound, the partial solution is left unfinished, and backtracking continues. If a new lower bound exceeds older ones, the latter are marked "dominated." Table 8-5, which shows how to record the computations, is much shorter than the lexical Table 8-4, but of course more computation went into it.

8.10.   LOGICAL PROBLEMS

Some of the most intriguing problems facing designers, or for that matter, detectives, jurors, physicians, or other troubleshooters, arise in human situations expressed in natural language instead of numbers. Statements like, "If the house was empty, then the butler did it," or "Appropriate penalties will be assessed if conditions (a) through (d) are not all met," supply constraints and objective functions in many realistic cases. There is a large literature of puzzles and brainteasers that reduce to finding the unique feasible solution to a set of complicated logical conditions like this one of Caliban's: "Mr. Forest's pigeon was killed by the cat owned by the lady who married the man whose pigeon was killed by Mrs. Field's cat." In real life there may not be enough statements of this sort to generate a unique solution, or worse, the statements may not be mutually consistent. In the former case, the multiplicity of feasible solutions can lead to an optimization problem if an appropriate objective function can be defined. In the latter, a reasonable objective is to find the solution minimizing the number of statements inconsistent with the others.

Such problems, once formulated, lend themselves to solution by lexical arithmetic. But to formulate them, one must first know *symbolic logic*, which gives precise and convenient ways of setting up logical statements. In symbolic logic, verbal statements are represented by latin letters a, b, c, etc. For example, a could stand for "the house was empty," while b might be "the butler did it." With each such statement is associated a bivalent variable, sometimes called a *boolean* variable in honor of George Boole, who studied such logical problems in the mid 19th century. Each bivalent variable is given the value unity if the statement is true but zero if it is false. Then for example, $a = 1$ means "the house was empty," whereas $a = 0$ means "the house was *not* empty." A number of logical relations between statements will now be presented.

Complementation is a function of a single bivalent variable, corresponding to the word "not" and symbolized by an overbar ($^-$). Thus "not a" is written $\bar{a}$ and called the *complement* of a. In the example, $\bar{a}$ means "the house was *not* empty." In ordinary arithmetic, $\bar{a} = 1 - a$ and $a = 1 - \bar{a}$.

Conjunction is a function of two variables corresponding to the word "and" and symbolized either by a dot ($\cdot$) or by placing the symbols for the statements next to each other as in ordinary algebraic multiplication. Thus "a and b" is written $a \cdot b$ or ab, meaning in the example that "the house was empty, *and* the butler did it." Other conjunctions such as "but" and "although" have the same logical function as "and," so a mystery writer can add suspense to ab by expressing it as, "Although the house was empty, the butler did it." Much of the challenge of logical problems is to see through the literary and the psychological relationships to the underlying logical ones. Table 8-6 gives the values of ab for each value of a and b individually. The marginal label a means $a = 1$, $\bar{a}$ means $\bar{a} = 1$ ($a = 0$), etc. The conjunction operation corresponds to ordinary algebraic multiplication.

Table 8-6   Conjunction $a \cdot b$

	b	$\bar{b}$
a	1	0
$\bar{a}$	0	0

Inclusive disjunction is a function of two variables that corresponds in English to the compound word *and/or* and whose symbol is v, sometimes called "cup" by mathematicians. Thus avb could be read "a and/or b," or in the example it could be stated "either the house was empty or the butler did it, or both." Notice that the English word "or" is ambiguous, for it does not make clear if exactly one of the possibilities is meant, or if both of them are allowed. In logic, the adjective *inclusive* specifies that both can occur. Table 8-7 gives the values of inclusive disjunction avb.

Table 8-7 Inclusive Disjunction $a \vee b$

	b	$\bar{b}$
a	1	1
$\bar{a}$	1	0

In ordinary algebra,

$$a \vee b = a + b - ab$$

which, being nonlinear, complicated Boole's attempts to solve logical problems with ordinary arithmetic. For present purposes it is easier to use symbolic logic directly than ordinary arithmetic.

Exclusive disjunction is a function of two variables corresponding to the verbal expressions "exactly one" or "either...or..., but not both." Symbolized by $\oplus$, it is called "symmetric difference" or "sum modulo 2" by mathematicians, but here it will be regarded merely as a shorthand symbol called "exc" for short. For example, $a \oplus b$ could be expressed as "either the house was empty, or the butler did it, but not both." See how poorly this common situation is expressed in English; this is why the symbolic form is so useful. It is interesting to see how it relates to the operations both of symbolic logic and of ordinary arithmetic.

$$a \oplus b \equiv a\bar{b} \vee \bar{a}b = a(1-b) \vee (1-a)b = (a-ab) \vee (b-ab)$$
$$= (a-ab) + (b-ab) - (a-ab)(b-ab)$$
$$= a + b - 2ab - ab + a^2b + ab^2 - a^2b^2$$

But since a and b are bivalent, $a^2 = a$ and $b^2 = b$, so

$$a \oplus b = a + b - 2ab$$

Table 8-8 gives the values of exclusive disjunction. In a given situation the ambiguity of the English word "or" may require careful examination of the context in which it occurs to determine which kind of disjunction is intended. When in doubt, ask questions.

Table 8-8 Exclusive Disjunction $a \oplus b$

	b	$\bar{b}$
a	0	1
$\bar{a}$	1	0

Implication is a function of two variables corresponding to "implies" or "if..., then..." in ordinary english, or at least in the more precise language of mathematicians. Its symbol is →. Thus $a \rightarrow b$ can be interpreted in the example as "If the house was empty, then the butler did it." In mathematics a is called the *hypothesis* and b the *conclusion*, and an implication is considered false only when the hypothesis is true but the conclusion is false. Thus the implication is regarded as true whenever the hypothesis is false, no matter what the conclusion, which would certainly seem bizarre to an engineer. This confusing convention will be justified when the next logical operation, equivalence, is discussed. Meanwhile, simply view an implication having an untrue hypothesis as true, but irrelevant. Table 8-9 gives the values of implication, and some expressions in terms of other logical and arithmetic operations follow.

$$a \rightarrow b = ab \vee \bar{a}b \vee \bar{a}\bar{b} = ab \vee \bar{a}(b \vee \bar{b})$$
$$= ab \vee \bar{a}$$
$$= ab + \bar{a} - ab\bar{a} = ab + \bar{a}$$
$$= 1 - a + ab$$

Table 8-9   Implication $a \rightarrow b$

	b	$\bar{b}$
a	1	0
$\bar{a}$	1	1

Equivalence is a function of two variables corresponding to "equals," "is the same as" or more cryptically, "if and only if." Its symbol is ↔. Thus in the example, $a \leftrightarrow b$ can be read, "The house was empty if and only if the butler did it," which wouldn't win any literary awards. Note that the mathematical jargon "if and only if" is itself an abbreviation of a more complicated statement which in the example would be, "The house's emptiness implies that the butler did it, and *vice versa* (i.e., if the butler did it, then the house was empty)." Thus equivalence can be expressed in terms of implication and conjunction as $a \leftrightarrow b = (a \rightarrow b)(b \rightarrow a)$. It is to make this true that an implication is considered true when its hypothesis is false. In terms of inclusive disjunction,

$$a \leftrightarrow b = (a \rightarrow b)(b \rightarrow a) = (ab \vee \bar{a})(ab \vee \bar{b}) = ab \vee ab\bar{b} \vee \bar{a}ab \vee \bar{a}\bar{b}$$
$$= ab \vee \bar{a}\bar{b}$$

Thus two statements are equivalent if either both of them are true or both

Table 8-10   Equivalence $a \leftrightarrow b$

	b	$\bar{b}$
a	1	0
$\bar{a}$	0	1

are false, as shown in Table 8-10. In ordinary algebra

$$a \leftrightarrow b = ab \vee \bar{a}\bar{b} = ab + \bar{a}\bar{b} - (ab)(\bar{a}\bar{b}) = ab + \bar{a}\bar{b} = ab + (1-a)(1-b)$$
$$= 1 - a - b + 2ab$$

Logical expressions can become quite complicated if not simplified. Here are some of the most widely used simplifying rules, proven by direct manipulation of the tables, called *Karnaugh maps*, defining the various operations. In these maps, widely employed in the design of computer circuit logic, each operation is performed in every cell of the combined map. For example, when elements of the map for a are multiplied with corresponding elements of the map for ab, the resulting map is identical with that for ab.

$$\begin{matrix} 1 & 1 \cdot 1 & 0 \\ 0 & 0 & 0 & 0 \end{matrix} = \begin{matrix} 1 \cdot 1 & 1 \cdot 0 \\ 0 \cdot 0 & 0 \cdot 0 \end{matrix} = \begin{matrix} 1 & 0 \\ 0 & 0 \end{matrix}$$

This proves the *Conjunctive Absorption Law*: $a(ab) = ab$
The *Disjunctive Absorption Law* is $a \vee ab = a$

Proof. $\begin{matrix} 1 & 1 \\ 0 & 0 \end{matrix} \vee \begin{matrix} 1 & 0 \\ 0 & 0 \end{matrix} = \begin{matrix} 1 & 1 \\ 0 & 0 \end{matrix}$

The *Cancellation Law* is $\bar{a} \vee ab = \bar{a} \vee b$

Proof. $\begin{matrix} 0 & 0 \\ 1 & 1 \end{matrix} \vee \begin{matrix} 1 & 0 \\ 0 & 0 \end{matrix} = \begin{matrix} 1 & 0 \\ 1 & 1 \end{matrix}$

The *Complementation (De Morgan) Laws* (1) $\overline{(a \vee b)} = \bar{a} \cdot \bar{b}$

Proof. $\dfrac{\overline{(1 \vee 1)} \quad \overline{(1 \vee 0)}}{\overline{(0 \vee 1)} \quad \overline{(0 \vee 0)}} = \begin{matrix} \bar{1} & \bar{1} \\ \bar{0} & \bar{0} \end{matrix} \cdot \begin{matrix} \bar{1} & \bar{0} \\ \bar{1} & \bar{0} \end{matrix}$

$$(2) \quad \overline{(a \cdot b)} = \bar{a} \vee \bar{b}$$

Proof.   Left as an exercise.

Symbolic logic is worth knowing if only for clarifying verbal complexities. But combined with lexical arithmetic in the next section, it becomes a powerful way to generate good designs even when verbal objectives conflict.

8.11. PRECEDENCE IN DESIGN

Many considerations, sometimes conflicting, go into a design. When they are expressed in natural language instead of numerically, they often can still be written precisely with symbolic logic. Then if an order can be established among the various goals, lexical arithmetic can be modified to generate designs tending to satisfy the objectives with the highest values, even though these values are not numerical.

This concept can be illustrated by a far-fetched example. Imagine choosing from among seven optional elements to be added to a device. In alphabetical order the elements are: axle, beam, crank, driveshaft, eccentric, flange, and guard. Each is represented by its initial letter, e.g., a for axle, etc., so that a means the device will have an axle, whereas $\bar{a}$ (or $a = 0$) means it will not, etc. There are $2^7 = 128$ possible designs to choose from.

The choices are restricted by needs for (1) reliability, (2) safety, (3) low cost, (4) marketability, (5) aesthetic appeal, and (6) low environmental impact. The most important of these for the device in question is high reliability, which would require that at least one of the first four components be present. This consideration can be expressed as the logical function $a \vee b \vee c \vee d$, which is 1 when the reliability condition is satisfied and 0 when it is not. Another form involving an ordinary algebraic inequality is $a + b + c + d \geqslant 1$.

The secondmost important considerations involve safety, and there are two of them. The first is that the device have either a crank and a driveshaft, or else an eccentric, both possibilities being allowed. This is expressed logically by $cd \vee e$. The second safety condition is that if a guard is present, then there must be a crank but not a driveshaft. In logical symbols, $g \rightarrow c\bar{d}$, which is also equivalent to $\bar{g} \vee c\bar{d}$, an example of a *disjunctive* form that will turn out to be more convenient than the original one for the method to be developed. Both safety restrictions are regarded as equally important, and an ideal design would satisfy both.

Third in importance is low cost, achievable if none of the first four elements are used. This could be expressed by the equations $a = b = c = d = 0$, or by the disjunctive form $a \vee b \vee c \vee d = 0$. But the method to be used requires a *characteristic function*, defined as a function taking on the value unity if and only if this condition is met. Here a convenient characteristic function is $\bar{a}\bar{b}\bar{c}\bar{d}$, obtained by complementing the disjunctive expression.

Fourth, there are three marketability considerations, all of equal importance. The first is that if the device has a crank and an eccentric but not a beam, then it must have a flange; in symbols, $\overline{bce} \to f$. The disjunctive form of this characteristic function would be $b \vee \bar{c} \vee \bar{e} \vee f$, but this is not used because it has more than two terms. The second marketability consideration is that having a guard but not an axle is equivalent to having a driveshaft but not a crank. Here the characteristic function $\bar{a}g \leftrightarrow \bar{c}d$ is preferred to the disjunctive form, which would have more than two terms. The third marketing consideration is that if there is no flange, there should be neither beam nor eccentric; $\bar{f} \to \bar{b}\bar{e}$, or in disjunctive form, $f \vee \bar{b}\bar{e}$.

A fifth type of objective is aesthetic. The device will definitely look better if there is no axle, eccentric, or guard whenever there is a crank. Notice this verbal way of expressing an implication, whose characteristic function is $c \to \bar{a}\bar{e}\bar{g}$ or $\bar{c} \vee \bar{a}\bar{e}\bar{g}$.

Finally, the environmental impact of the device is lessened if there are at least two elements chosen from either the first four or the last three items. In mixed logical and arithmetic symbols, the characteristic function is $(a+b+c+d \geqslant 2) \vee (e+f+g \geqslant 2)$.

The nine characteristic functions for the restrictions are now listed, with low numbers given to those with high precedence. Those like the safety considerations that have equal value are given the same number, but with a letter following to distinguish between them, e.g., 2a and 2b. The particular form of characteristic function selected is preferably a single conjunction or pair of conjunctions joined by the inclusive disjunction operator $\vee$, the conjunction with the fewer variables being written first to the left. Thus $g \to c\bar{d}$ is written $\bar{g} \vee c\bar{d}$. When such a disjunctive form would have more than two conjunction terms, the more compact binary form is used. That is, $\overline{bce} \to f$ is preferable to $b \vee \bar{c} \vee \bar{e} \vee f$, and $\bar{a}g \leftrightarrow \bar{c}d$ is better than $\bar{a}g\bar{c}d \vee (\bar{a}g)(\bar{c}d)$. Ordinary algebraic statements such as $a+b+c+d \geqslant 2$ are written when the equivalent disjunctive expression has more than two conjunction terms, as in $ab \vee ac \vee ad \vee bc \vee bd \vee cd$.

The logical conditions

1.	$a \vee b \vee c \vee d$	reliability
2a.	$e \vee cd$	safety
2b.	$\bar{g} \vee c\bar{d}$	
3.	$\bar{a}\bar{b}\bar{c}\bar{d}$	cost
4a.	$\overline{bce} \to f$	
4b.	$\bar{a}g \leftrightarrow \bar{c}d$	marketability
4c.	$f \vee \bar{b}\bar{e}$	
5.	$\bar{c} \vee \bar{a}\bar{e}\bar{g}$	aesthetic
6.	$(e+f+g \geqslant 2) \vee (a+b+c+d \geqslant 2)$	environmental impact

8.12. SOLUTION BY LEXICAL ARITHMETIC

This logical design problem will now be solved by lexical arithmetic, extended to encompass this new nonlinear situation, certainly more complicated than the linear tolerance problem. Table 8-11 shows the computations, which will be described line by line. New rules and conventions will be explained as needed. The procedure is intended to make the solution of logical optimization problems almost as straightforward as grammar school arithmetic.

The method involves implicit enumeration on the design variables a through g, with the order of enumeration guided by the precedence order of the conditions. One tries to satisfy high-priority conditions first, using good solutions to end searches that cannot generate better ones. As assignments of variables are made, all simplifications that result are recorded. A history of the variables assigned is arranged at the right of the table in a manner that facilitates backtracking. As solutions are generated, they are written to the far right in terms of conditions satisfied so that lexicographic comparison is easy.

Begin by setting reliability condition 1 ($a \lor b \lor c \lor d$) to unity. In the example, many assignments would do this, but rather than pick any one of them, let condition 1 be imposed as a constraint that all later assignments must satisfy. The asterisk (*) shows that condition 1 is temporarily considered a constraint and must be checked whenever an assignment is made. Although no variable has yet been assigned, this completes line 1. The zero on the line will be generated and explained later.

Next examine safety condition 2, which has two parts. The first, 2a, can be satisfied with the simple assignment $e = 1$, so e is written in the first solution column on line 2. This neither violates nor satisfies the imposed constraint, whose column is left blank, and a 1 is entered on line 2 under condition 2a, now satisfied. For four other conditions, simplifications result which are recorded on line 2. For example, marketing condition 4a simplifies from $bce \to f$ to $bc \to f$, and the first environmental condition 6 becomes $f + g \geqslant 1$, which can be written in logical form as $f \lor g$.

Line 3 is generated by choosing $\bar{g}$, the simplest assignment satisfying the next open condition 2b, with the results indicated. Condition 4b becomes $0 \leftrightarrow \bar{c}d$, or $\bar{c}d = 0$, which becomes $c \lor \bar{d}$ when complemented to get a characteristic function. The new assignment $\bar{g}$ is placed on line three just to the right of e, the previous assignment. Any blanks on the line represent expressions that, being unchanged from lines above, need not be recopied.

Line 4 comes from forcing the next condition (3) to unity, which occurs when $\bar{a}\bar{b}\bar{c}\bar{d} = 1$. This multivariable expression involves only conjunctions, so simultaneously set $\bar{a} = \bar{b} = \bar{c} = \bar{d} = 1$, and record $\bar{a}\bar{b}\bar{c}\bar{d}$ just to the right of $\bar{g}$ in

Reliability 1	Safety 2a	Safety 2b	Cost 3	Marketability 4a	Marketability 4b	Marketability 4c	Aesthetic 5	Environmental 6	Assignments	Solutions	Line
$a \lor b \lor c \lor d$	$e \lor c \lor d$	$\bar{g} \lor c \lor d$	$\bar{a}\bar{b}\bar{c}\bar{d}$	$\bar{b}c \to f$	$\bar{a}g \leftrightarrow cd$	$f \lor \bar{b}\bar{e}$	$\bar{c} \lor \bar{a} \lor e \lor \bar{g}$	$(e+f+g \geq 2) \lor (a+b+c+d \geq 2)$			
$(a \lor b \lor c \lor d)^*$			0						1*		1
	1			$\bar{b}c \to f$		f	$\bar{c}$	$f \lor g \lor (a+b+c+d \geq 2)$	e		2
		1			$c \lor \bar{d}$			$f \lor (a+b+c+d \geq 2)$	$\bar{g}$		3
0^*			1						$\bar{a}\bar{b}\bar{c}\bar{d}$		4
1^*				$b \lor f$	1		0	$f \lor a \lor b \lor d$	c		5
					1	1		1	$e\bar{g}cf$	$\bar{3}, \bar{5}(d)$	6
$(a \lor b \lor d)^*$	1		0	1	$\bar{d}$	f	1	$f \lor (a+b+d \geq 2)$	$e\bar{g}\bar{c}$	$\bar{3}$	7
$(a \lor b)^*$				1	1	1		$f \lor a \lor b \lor e \lor k \lor g \lor c \lor d$	$\bar{d}$		8
1^*								1	$e\bar{g}\bar{c}\bar{d}f(a \lor b)$	$\bar{3}$	9
1^*	1		0	1	0		$\bar{c}$	1	$e\bar{g}\bar{c}d$	$\bar{3}, \overline{4b}(d)$	10
$(a \lor b \lor c \lor d)^*$		$c\bar{d}$	0	$\bar{b}c \to f$	$\bar{a} \leftrightarrow cd$	f	$\bar{c}$	1	eg		11
1^*	1			$b \lor f$	a		0		$(c\bar{d})$	$\bar{3}, \bar{5}(d)$	12
$(a \lor b \lor c \lor d)^*$	cd	$\bar{g} \lor c \lor d$	0	1	$\bar{a}g \leftrightarrow cd$	$f \lor \bar{b}$	$\bar{c} \lor \bar{a}\bar{g}$	$f \lor g \lor (a+b+c+d \geq 2)$	$\bar{e}$	$\bar{3}$	13
1^*	1	$\bar{g}$			$a \lor \bar{g}$		$\bar{a}\bar{g}$	1	(cd)		14
	1	1			1	1	1	1	$\bar{e}cd\bar{g}(f \lor \bar{b})\bar{a}$	$\bar{3}$	15
0^*									$\bar{1}^*$	$\bar{1}(d)$	16

the solution columns of line 4. However, this violates the imposed constraint $a \vee b \vee c \vee d$, shown by the 0 in that column, so the entire line is crossed out to show its infeasibility. The impossibility of satisfying cost condition 3 as long as reliability condition 1 holds is recorded by a zero under condition 3 on line 1, where the constraint was imposed. Line 4 is now regarded as inoperable, so all expressions above it carry down to the lines below.

Next in priority are the three marketing conditions, of which both 4b and 4c are satisfiable by fixing any of the three variables c, $\bar{d}$, or f. Arbitrarily take the first, and set $c = 1$, with the results shown on line 5. Notice that $\bar{b}c \rightarrow f$ becomes $\bar{b} \rightarrow f$, written in disjunctive form as $b \vee f$. Aesthetic condition 5 is violated, as shown by the 0 in that column. The imposed constraint 1 is finally satisfied.

The assignment of $f = 1$ now takes care of all three remaining restrictions 4a, 4b, and 6, as shown on line 6. Three variables—a, b, and d—have become irrelevant in that their values do not affect the outcome. Thus the solution $e\bar{g}cf$ actually represents $2^3 = 8$ separate designs, for this is the number of ways that zeroes or ones can be assigned to the three unspecified variables. The lexicographic form of the solution would be (1; 2a, 2b: 4; 6), but the complement $(\overline{3, 5})$, being shorter, is written to the far right of line 6.

Since the assignment on line 6 completed a solution, its complement would violate some of the conditions previously satisfied and be clearly dominated. Backtrack by changing the assignment in preceding line 5 from c to $\bar{c}$. The earlier assignments $e\bar{g}$ remain in force, so line 7 is generated by applying $\bar{c}$ to line 3, where the partial solution was $e\bar{g}$. All expressions obtained must be written down, since they are derived from line 3 rather than line 6 immediately above.

Line 8 comes from assigning $d = 0$ to satisfy the next condition in order. Then it becomes clear that all remaining conditions can be satisfied by any of the three assignments whose characteristic function is $f(a \vee b)$, written to the right of line 9. Since only cost condition 3 is violated by this solution, it dominates the one on line 6.

This is in fact the best attainable, for the only way cost condition 3 can be satisfied is to violate reliability condition 1, a solution which would of course be dominated. Hence computation can stop here if the designer subscribes to the principle of *satis quod sufficit*. It is not unreasonable, however, to see if any other designs can be as good as the three in hand, so for this reason, and to gain more practice with nonlinear lexical arithmetic, let us generate all alternate optima.

The complement of the last assignment, as always, is dominated, so backtracking proceeds by complementing the assignment preceding. This sets $d = 1$ in the expressions of line 7, as shown on line 10. The violation of 4b rules out this partial solution.

Seven more lines are needed to complete the backtracking, the three new optimal designs $\bar{e}cd\bar{g}(f \vee \bar{b})\bar{a}$ being discovered on line 15. Notice that line 16 backtracks not on a design variable, but rather on the restriction 1 imposed on line 1. The first 15 lines were carried out under this artificially imposed constraint, so the delayed alternative must be examined before implicit enumeration is complete. In this case the alternative is clearly dominated, so the search ends immediately.

The optimal design comes in six models, for which the characteristic function can be written $\bar{g}(\bar{c}d\bar{e}(a \vee b) \vee \bar{a}cd\bar{e}(b \vee f))$, with no further simplification possible. None of them have guards. Three of the models have an eccentric but no crank or drive. One of these also has an axle, another has a beam, and the third has both. The other three models are without axle or eccentric, but they have both crank and drive. Two of these are beamless, the flange being optional, while the third is beamed and flangeless. Speculations on the nature of this outrageous device are left to the imaginative reader.

Those who have struggled through the example can now appreciate the following remarks. The procedure is first of all a scorekeeping system, the hard part about logical problems being to keep track of what is happening. Beyond this, the method is effective because it allows branching on the restrictions themselves as well as on the variables. And it provides a rational way to decide which to assign next. A final feature to notice is that constraints cause no additional difficulty. In the example, the first restriction was in fact treated as a constraint throughout the computation.

Armed with lexical arithmetic, the engineer can begin to analyze some aspects of design which, being verbal rather than numerical, have in the past been relegated to the "Art" of design. Let's see if "Science" can get any further.

8.13. CONCLUDING SUMMARY

This chapter has shown that dealing with indivisible quantities is not as difficult as it may first have seemed. The bounds needed for eliminating mediocre designs by implicit enumeration are often generated readily by the methods of the early chapters. Choosing among discrete alternatives, as in the tolerance allocation problem, can be done systematically and rigorously with lexical arithmetic. Combined with symbolic logic, lexical arithmetic can master bewildering verbal problems arising from laws, contracts, and customer requirements. Discrete problems seem hard because conventional calculus cannot be used on them. What makes them easy in reality is that only a finite number of possibilities need be examined and the computations often involve only addition and subtraction. The main concern is keeping track of what happens.

Implicit enumeration is in itself worth learning by engineers, whether or not they have design problems, for in any study it quickly eliminates mediocre ideas. Symbolic logic is also valuable for troubleshooters of all sorts, be they engineers, physicians, or private eyes.

Lexical arithmetic holds great promise for engineering analysis. It provides a simple, rapid way to allocate limited resources, whether they are machine tolerances or construction capital, among competing processes or projects. This tabular scheme can be used equally well when costs or profits are not well known numerically but can be arranged in relative order. Lexical arithmetic selects, from among many possibilities, the few locally dominant designs that the designer may wish to study further. Thus the procedure is useful even when cost information is imprecise or unavailable. Combined with symbolic logic, it may generate promising designs that might well be overlooked in a less systematic search. This is one way inventions are discovered.

NOTES AND REFERENCES

Foundations of Optimization has more on Fibonacci search and its extensions by Avriel, which Beamer and Shapiro have taken farther.

Many fundamental developments in discrete optimization come from former students of the late Professor Moisil of Bucharest; namely, Balas, Hammer, and Rudeanu. Balas introduced backtracking into integer programming, and pseudoboolean programming was developed by Hammer and Rudeanu.

Lexical arithmetic has origins in the work of Barthès on simple methods for implicit enumeration.

The myriad computer oriented integer programming methods are covered well by Garfinkel and Nemhauser.

Avriel, M., and D. J. Wilde, "Optimal Search for a Maximum with Sequences of Simultaneous Function Evaluations," *Manage. Sci.*, **12**, 722–731 (May, 1966).

Balas, E., "An Additive Algorithm for Solving Linear Programs with Zero-One Variables," *Oper. Res.*, **13**, 517–546 (1965).

Barthès, J. P., *Branching Methods in Combinatorial Optimization*, Ph.D. dissertation, Stanford, 1973.

Beamer, J. H., and D. J. Wilde, "Minimax optimization of a unimodal function by variable block derivative search with time delay," *J. Combinatorial Theory*, **10**, 2 (Mar. 1971) 160–173.

Boole, G., *An Investigation of the Laws of Thought*, Dover, New York, 1854.

Garfinkel and Nemhauser, *op. cit.* Chapter 1.

Hammer and Rudeanu, *op. cit.* Chapter 1.

Kiefer, J., "Sequential Minimax Search for a Maximum," *Proc. Amer. Math. Soc.*, **4** (1953) 502.

Shapiro, R. J., and D. J. Wilde, "Optimal Minimax Search with Unequal Block Sizes," Tech. Rpt. 74–16, Dept. Opns. Res., Stanford, (Sep 1974).

Wilde and Beightler, *op. cit.* Chapter 1.

EXERCISES

8-1. Find a satisfactory design for this slightly modified version of the cofferdam problem introduced in Sec. 2.6:

$$\min c = 168st + 3660t + 4.57(10^6)stn + 50000h^{-1}$$

subject to:

$$s \geq 1.0425t \qquad 0.00035ft \leq 1 \qquad t \geq h + 41.5 \qquad n = 800f^{-1}$$

with n, the number of cells, required to be an integer, as well as s and f (Sec. 8.1).

8-2. Suppose that in the merchant fleet design problem of Sec. 4.8, the length l, beam b, and height h must all be integers, as well as the number of ships. Find a satisfactory design (Sec. 8.1).

*8-3. Indicate which variables in your design project should in practice be confined to standard sizes and list what these are. Is your rounded design still satisfactory? If not, incorporate branch and bound into your design procedure (Sec. 8.1).

8-4. The cost of an endothermic chemical reactor (Sec. 7.1) is proportional to $100(T/1000)^4 + \exp(5000T^{-1})$, where T is the absolute temperature in °K. Assuming the minimizing temperature is between 1000 and 1500°K, and that only multiples of 10°K can be considered, find the minimizing temperature by Fibonacci search (Sec. 8.2).

8-5. Solve the tolerance assignment problem of Sec. 8.3, changing the total tolerance from 24 to 23 mils (Secs. 8.4–8.6).

8-6. Change the total tolerance of Sec. 8.3 to (a) 25 mils: (b) 30 mils. Use lexical arithmetic to solve the problems resulting (Secs. 8.7 and 8.8).

8-7. Apply the objective function bounding method of Sec. 8.9 to the preceding problem.

8-8. Replace inclusive by exclusive disjunction in the safety and aesthetic conditions of Sec. 8.11 and solve the new problem resulting.

8-9. The Curious Case of the Moldavian Prince.

Never have I seen a case with such high ranking suspects, Holmes. Imagine, an Archduke, a Bishop, a Count, a Duchess, and even an English Earl.

*Project problem.

All cousins of the murdered Prince of Moldavia, Watson. I am absolutely certain that at least one of them committed the atrocity.

I wish we could be as positive about the other deductions, made by our field operatives: Archer, Bogart, Hammer, Marlowe, Spade, and Tracy.

Excellent souls, Watson, although prone to error.

The most plausible theory is Archer's. He concludes that the Count and the Earl could have done it only if the Archduke, their liege lord, was involved.

I agree on its plausibility, Watson. Next most believable is Bogart's side-of-the-mouth observation that the high heeled footprints on the Prince's chest could only have been made by the Duchess or the Count.

I never did like the looks of that Trans-whatever-he-is, Holmes. I suppose the next most credible hunch is Marlowe's. The Triconopoly cigar ash in the Salem box suggests to him that at least two of the men were involved.

Good thinking, but really less believable than Hammer's scandalous report about the Bishop and the Duchess. One of them never commits anything, even murder, without the other being present.

Cherchez la femme, Holmes. That leaves the two *least* plausible ideas.

Of the two, I prefer Spade's. The Maltese Falcon sweatshirt certainly indicates that at least one of the male aristocrats was involved, Watson.

That leaves us with Tracy's report.

Yes, Watson. Incompetent though he be, there is no denying that the Count and the Duchess have avoided each other since their divorce.

What are we to do then, Holmes?

Elementary, my dear Watson. We increase surveillance of those suspects whose guilt makes the six ordered statements lexically maximal without violating our absolute certainty that at least one of them did it. Pass me the graph paper and the seven percent solution.

CHAPTER **9**

Perspectives

*Learning without thought is labor lost; thought
without learning is perilous.*

Confucius (551–479 B.C.)

Now is the time to reflect on what these seven technical chapters have
said. Ten major themes have been developed, 16 examples solved, and four
branches of optimization theory pertinent to design presented and extended in six new directions. All this will be summarized and reviewed in
this final chapter.

Once this has been done, locating the book within the context of design
on the one hand and optimization theory on the other will be straightforward. Related work will be described and research areas indicated. This
will guide the reader to complementary topics in both design and optimization if wider study of either discipline seems attractive.

9.1. MAJOR THEMES

Ten major themes run through this work. Each represents a strategy, an
approach, or a principle that, once familiar, becomes a useful habit for
designers facing a new problem. The 10 are, in rough order of their
appearance: (1) partial optimization, (2) monotonicity analysis, (3) bounding function construction, (4) conditional design, (5) stationarity (and
orthogonality) analysis, (6) logarithmic (and semilogarithmic) derivatives,
(7) scaling, (8) reversed inequality elimination, (9) transcendental programming, and (10) lexical arithmetic.

257

Three of these are simplification strategies for reducing the complexity of the original problem, sometimes all the way to a solution. These are partial optimization, which cuts down the number of variables directly, and monotonicity and stationarity analysis, which determine constraint activity for subsequent elimination of variables and constraints. Their mutual application often leads to a conditional-design procedure, if not to a single formula, giving the optimal design after a few closed-form computations with the design parameters.

Scaling and reversed inequality elimination are also preliminary approaches intended not so much for simplification as for clarity. A reversed inequality often signals a local minimum, easily overlooked because of its association with a simpler equipment configuration, warning the designer to construct global bounds in seeking the global optimum. Scaling reveals the distribution among the system elements, often exposing numerical errors in the initial formulation. This not only helps the designer neglect small items to focus on important ones, but also suggests dual weightings for condensation and bound construction.

Bounding functions can greatly speed generation of a satisfactory design numerically indistinguishable from the theoretical optimum. This attitude of *satis quod sufficit* is made mathematically rigorous to reduce unproductive computing and, perhaps more important, to keep the designer from overlooking any novel but unsuspected global optimum far from the base case. Bounding tames engineering nonlinearities, eliminates myriad discrete possibilities, and in the form of lexical arithmetic, even copes effectively with logical complexities.

The logarithmic derivative and its semilogarithmic cousin make available first and second derivatives, not just for numerical searches, but for simple derivation and interpretation of stationarity and orthogonality conditions as well. They show how algebraic variables can disappear from transcendental problems and what is needed to take care of transcendentality.

The 10 themes recur and interact throughout the book. Mastery of them can only sharpen the designer's skill.

9.2. EXAMPLES

The 16 examples apply and illustrate theoretical principles of general interest to all designers. Hence their content is deliberately unspecialized to make them physically comprehensible to any broadly educated engineer. Yet five of them should be especially interesting to designers with a mechanical background—machine designers and naval architects, as well

as industrial and aerospace engineers. These examples are the two tolerance specification problems, the hydraulic cylinder, the merchant fleet,
and the gadget design illustration at the end of the preceding chapter. Four
more examples fall within the province of chemical engineering and its
offshoots, process and nuclear engineering: the fertilizer plant, the reactor
cooler, the chemical reactor, and the gaseous diffusion plant. Civil engineers would probably be most interested in the cofferdam, gravel sled,
and irrigation reservoir examples. Of the four examples remaining, the
pressure vessel, ammonia storage, and desalination plant all involve process technology important to both chemical and mechanical engineers.
Pipeline problems like that in the example could be assigned to anyone
familiar with hydraulics and fluid mechanics, certainly any qualified civil,
mechanical, or chemical engineer in the design office. Only the electrical
engineers have been slighted, but they can find good examples in Zener's
book.

The ease with which the original problems were reduced to only one or
two variables may make the reader forget how complicated were
the original statements. Most problems had at least four variables in the
beginning, and the fleet design example had nine. More important, the
lack of linearity or even convexity put these problems, especially
the multimodal ones of Chapter 6, out of reach of existing computerized
nonlinear programming methods.

9.3. CONTRIBUTIONS TO OPTIMIZATION THEORY

Although written for designers, this book could not avoid expanding the
scope of optimization theory in ways that would interest applied mathematicians. When existing theory could not cope with the peculiarities of
design problems, new ideas had to be developed.

Present optimization theory shown pertinent to design included geometric programming, the theory of quadratic forms, and the stabilized N-R
method. The extensions needed to handle signomial and transcendental
functions involved logarithmic derivatives, reversed inequalities, and secant and tangent bounding functions. Also lexical arithmetic can be
considered a designer's adaptation of the existing theory of pseudoboolean
programming crossed with branch and bound.

From its conception, geometric programming was intended to be an aid
to designers. It is perfect for the power functions of engineering, and the
elegance with which it solves the zero-degree-of-difficulty case is unforget-
table. Negative signs, whether from profit considerations or reversed
conservation inequalities, require either signomial extensions based on

semilog derivatives, or else the secant and tangent bounding functions to secure the global optimum. These extensions in turn are based on more existing theory, that of quadratic forms and of the N-R method. The new transcendental theory is most powerful when circumstances permit geometric-programming-style bounds to be constructed.

Although developed to solve design examples, lexical arithmetic can solve operating problems just as easily. The discrete tolerance assignment problem could in fact be viewed as one in scheduling manufacturing processes. And the adaptation of lexical arithmetic to logical problems is ready made for troubleshooters.

There may well be nonlinear, nonconvex, multimodal operations research problems solvable by these methods, inspired though they be by engineering design. Certainly the mathematics won't mind.

9.4. RELATED TOPICS IN OPTIMIZATION

Presented here has been a particular collection of optimization techniques capable of solving problems peculiar to design. Those wishing to broaden their competence in optimization beyond that directly applicable to design would do well to examine linear and dynamic programming, direct search, and penalty functions. Control and probability theory are also good background.

Linear programming, dealing with linear objective and constraints, is of course one of the cornerstones of operations research. Although not very useful for the highly nonlinear problems with few variables studied here, it seems to be the main technique now available for studies, preliminary at least, of systems with many variables. After all, in the design of large manufacturing complexes, one cannot really avoid operating problems, which usually introduce enough linearity to make linear approximations appropriate. Present theory for optimizing large systems is heavily influenced by linear programming and its particular duality theory, quite different from that of geometric programming. Some day the parametric analysis of this book may be extended to the assembly of components into large systems, but for now the best preliminary approach is still linear programming.

Dynamic programming, the forerunner of partial optimization, has been used a little by designers, but the applications were not widespread. Its main power is for systems of components in series coupled by no more than two variables. Loops and branches complicate the problem to the point where it is all too easy to get the wrong answer, but careful analysis is needed for the methods of this book also. As the tolerance assignment problem of Chapter 8 demonstrated, many dynamic programming problems involving discrete variables can be solved more easily by lexical arithmetic. Dynamic programming has been applied to automatic control optimization problems, also soluble by the maximum principle, which can handle many more coupling variables. Competence in optimal control theory is certainly valuable to any designer with dynamic, that is, time-varying problems.

There are also three approaches that, while giving little or no insight into a design problem, can at least generate a quick (possibly wrong) answer in a hurried situation. These are the penalty function, direct search, and differential optimization methods. The penalty function method of Fiacco and McCormack, with refinements by Lootsma, replaces a constrained optimization problem by an unconstrained one in which constraint violations are penalized economically in the units of the objective. This technique is valuable when there are many inequality constraints but no good idea which are active at the optimum. Followed by a more rigorous analysis of the constraints found active in this way, the dangers of convergence to the wrong local minimum can be reduced.

Because it is mathematically complicated, the penalty function usually requires direct search for its minimum, and there are many ways to do this besides using the N-R method described in this book. Once the power function structure, with its easily computed first and second derivatives, has been lost, the other direct-search methods become important, so the designer should have at least a nodding acquaintance with them.

Differential optimization theory shows how to generate derivatives for N-R moves before constraint activity has been determined. Material on this subject, almost included in this book, is in the second edition of *Foundations of Optimization*, also a good engineers' reference on linear, dynamic, and geometric programming. The most important implication of differential optimization for design may be its potential value in large problems involving assemblies of components, a topic on which research has only begun.

A key matter not considered in this book is uncertainty in design. The design parameters that govern the optimal design may not be known constants as assumed here. Instead they may be random variables, possibly expressed as probability distributions. Little work has been done on this problem in a design context. The research-minded designer may therefore wish to study probability and statistics to guide future applications of this new branch of decision making under uncertainty.

These topics—linear and dynamic programming, control theory, penalty functions, direct search, differential optimization, and probability theory—should be considered by the designer as nice, although not essential, to know. They are less important than those in this book, which have been selected to see the engineer through most practical problems in optimal design.

9.5. THE PLACE OF OPTIMAL DESIGN

Optimal design is just part of design, not the whole. A good designer must know many other things and, most important, be able to create with them.

Some of the knowledge a designer needs is now taught in engineering colleges. Basic science, mathematics, and the fundamental engineering sciences for the specialized disciplines are standard components of any engineer's education. Other things must be learned on the job. Technological information such as terminology, codes, standards, and practices change too rapidly and are too specialized to be in a college curriculum, where there is hardly enough time to master the fundamentals. Trade journals, company manuals, vendors' catalogs, and technical monographs supply information of this sort, to say nothing of word of mouth communication.

Aside from engineering science and technology, there is other knowledge a designer needs that is teachable in an academic setting. Visual thinking and graphic communication are examples of such subjects that can be taught at any level in the curriculum. There is also a design course, usually at the senior level, that gives students the chance to integrate earlier material, learn how to research a specialized technology, work as part of a design team, and experience the joy and pain, if not of creation, then at least of synthesizing a new machine, process, or structure from myriad components. All these are, in the main, undergraduate activities.

At the postgraduate level there are possibilities for advanced design education aside from the obvious use of graduate courses in engineering science. Computer-aided design brings mathematics and computer graphics to bear on complicated geometry and analysis problems. Engineering case study, like similar courses in law and business schools, trains the designer to cope with the complexities of real engineering practice. A graduate design course gives valuable guided experience in designing advanced devices, structures, or processes, just as a systems design course prepares future project managers.

Here in this constellation of graduate design courses is where optimal design belongs. By requiring students to find, model, and solve their own individual design projects, it gives them the opportunity to integrate knowledge from the entire curriculum. They find how easy it is to overlook important costs and constraints needed to generate a sensible design. They learn the futility of excessive numerical perfectionism. And they grasp how economics and physics interact to shape an engineering device.

A course in optimal design is no substitute for an undergraduate design course, which should be where the student first encounters broad design questions. Good undergraduate design problems are usually constraint bound, with the instructor guiding the modeling and computation of revealing cases. Only after such an experience is a design student ready for the degrees of freedom and multiple optima of optimal design.

Of course there are many things in design that must be learned but cannot be taught. Experiential courses can simulate the realities of design in hopes that the students will teach themselves. But for the most part, universities can only take responsibility for the teachable part of design. Certainly teachable and eminently useful, optimal design is now ready to take its place among the graduate courses of engineering design.

From the Egyptian sage Amenemope (10th century B.C.) comes an exhortation: "Weary not thyself to seek for more." Good as this would be as an ending line for this book, a few more words make it into an excellent summary as well:

Weary not thyself seeking more than there is.

Author Index

265

Subject Index